RESEARCH IN MOLECULAR LASER PLASMAS
ISSLEDOVANIYA PLAZMY MOLEKULYARNYKH LAZEROV
ИССЛЕДОВАНИЯ ПЛАЗМЫ МОЛЕКУЛЯРНЫХ ЛАЗЕРОВ

The Lebedev Physics Institute Series

Editors: Academicians D. V. Skobel'tsyn and N. G. Basov

P. N. Lebedev Physics Institute, Academy of Sciences of the USSR

Recent Volumes in this Series

Proceedings (Trudy) of the P. N. Lebedev Physics Institute

Volume 78

RESEARCH IN MOLECULAR LASER PLASMAS

Edited by
N. G. Basov

P. N. Lebedev Physics Institute
Academy of Sciences of the USSR
Moscow, USSR

Translated from Russian by
Albin Tybulewicz

Editor, *Soviet Journal of Quantum Electronics*

CONSULTANTS BUREAU
NEW YORK AND LONDON

Library of Congress Cataloging in Publication Data

Main entry under title:

Research in molecular laser plasmas.

(Proceedings (Trudy) of the P. N. Levedev Physics Institute; v. 78)
Translation of Issledovaniiâ plazmy molekuliârnykh lazerov.
Includes bibliographical references and index.
1. Plasma (Ionized gases)—Addresses, essays, lectures. 2. Carbon dioxide lasers—
Addresses, essays, lectures. I. Basov, Nikolaĭ Gennadievich, 1922-   II. Series:
Akademiiâ nauk SSSR. Fizicheskiĭ Institut. Proceedings; v. 78.
QC1.A4114 vol 78   [QC718.15]   530'08s   [530.4'4]
ISBN 978-1-4684-1625-1   ISBN 978-1-4684-1623-7 (eBook)   76-25553
DOI 10.1007/978-1-4684-1623-7

The original Russian text was published by Nauka Press in Moscow in 1974 for the
Academy of Sciences of the USSR as Volume 78 of the Proceedings of the P. N.
Lebedev Physics Institute. This translation is published under an agreement with the
Copyright Agency of the USSR (VAAP).

PREFACE

This volume reports investigations which form part of a major series of theoretical and experimental studies being carried out in the Laboratory of Low-Temperature Plasma Optics at the Lebedev Physics Institute in Moscow. The papers give the results of systematic investigations of the chemical composition and of the electrical and optical properties of discharge plasmas, and also of populations of laser levels. Reliable and detailed information is given on the dissociation of carbon dioxide gas in discharges; the nature of the velocity distribution function, average energies, and densities of electrons; and populations and vibrational temperatures of molecules in cw CO_2 and CO lasers.

The material in this volume is intended for specialists in quantum electronics and low-temperature plasma diagnostics.

CONTENTS

INVESTIGATIONS OF PHYSICOCHEMICAL
PROPERTIES OF CO_2 LASER PLASMA*

V. N. Ochkin

The infrared absorption method was used in a study of the dissociation of CO_2 in the plasma of gas-discharge CO_2 lasers. The degree of dissociation in continuous-flow CO_2 lasers was considerable (reaching 60% or more) even at high flow rates. A calculation was made of the population inversion density allowing for dissociation and it was found to be strongly inhomogeneous along the discharge tube in continuous-flow systems because of the inhomogeneity of the chemical composition of the plasma. The electrode material and its treatment were of considerable importance in the case of sealed systems. Processes which governed the service life of a sealed CO_2 laser were identified. A method for the determination of the vibrational temperatures of N_2, $N_2 - CO_2$, and $N_2 - CO_2 - He$ gases in the ground electronic states was developed (and used). These temperatures were deduced from the relative intensities in the second positive system of nitrogen.

INTRODUCTION

In 1964, Patel et al. [1] achieved continuous stimulated emission of a large number of lines in vibration-rotation bands of the CO_2 molecule in the 10-μ range. The output power was only 1 mW. Similar results were reported almost simultaneously by French workers [2]. Soon after many investigators reported outstanding power outputs. For example, Roberts et al. [3] achieved a power of about 5 kW for an efficiency of ~25%. Such high-power carbon dioxide lasers are complex systems. On the other hand, CO_2 lasers used in the laboratory are relatively simple. As a rule, they consist of a running-water-cooled glass or quartz discharge tube, 1-2 m long and of several tens of millimeters in diameter. Such a tube is placed inside a Fabry—Perot resonator consisting of two plane or spherical mirrors; the transmission of light through one or both of these mirrors differs from zero. A glow discharge takes place in a mixture of gases containing CO_2. The discharge current (which may be direct or alternating) is 10-100 mA and the total voltage drop across the tube is 5-20 kV. The most widely used

---

* Candidate's dissertation, defended in April 1970 at the P. N. Lebedev Physics Institute, Academy of Sciences of the USSR, Moscow. The work was carried out under the direction of N. N. Sobolev and É. N. Lotkova.

mixture of gases is composed of CO_2, N_2, and He; its total pressure is several Torr. Typical output powers of such lasers amount to a few tens or hundreds of watts. Output power densities up to 9 W/cm^3 can be obtained by special cooling and at higher pressures (several tens of Torr); in this case, the total output power may reach 1 kW per 1 m of the discharge gap [4].

The high performance and extensive applications of the carbon dioxide laser have stimulated intensive investigations.*

Most of the early work on CO_2 lasers was purely empirical. A detailed identification of laser transitions was made, optimal conditions in the discharge were determined, resonator characteristics were optimized, etc. This resulted in a rapid increase in the output power of CO_2 lasers.

Much more difficult was the problem of the population inversion mechanism. The first attempt to explain the operation of a CO_2 laser was made by Patel [7] for the case of a discharge in a mixture of CO_2 and N_2 gases. He assumed that the upper laser level of the CO_2 molecule was populated by collisions with vibrationally excited nitrogen molecules whose vibrational levels were in good resonance with antisymmetric vibrations of the CO_2 molecule. He proved this hypothesis by the experimental data in which N_2 was excited separately and then mixed with CO_2 [8]. Patel suggested that the strong excitation of the molecular vibrations of N_2 was due to electron—ion and atom—atom recombination processes, as well as to cascade transitions from excited electronic states. However, Patel's theory failed to account for the high powers and efficiencies which were soon reached.

The undoubted practical and theoretical importance of the population inversion mechanism in the discharge plasma of the CO_2 laser resulted in several investigations of this subject.

The present paper belongs to an extensive series of theoretical and experimental investigations carried out in the Laboratory of Low-Temperature Plasma Optics of the Lebedev Physics Institute in Moscow.

The first theoretical papers of this series were written by Sobolev and Sokovikov, who suggested a direct electronic excitation of the molecular vibrations resulting in the population of the upper laser level [9]. This hypothesis was based on the experimental results of Schulz [10] on the probabilities of the excitation of molecular vibrations by electron impact and it soon gained general acceptance. An analysis of the experimental and theoretical material, however, led Sobolev and Sokovikov to the conclusion that the depopulation of the lower laser levels of CO_2 was due to collisions of CO_2 with atoms and molecules [11]. These physical considerations were used in [12] to calculate the population inversion in a CO_2 laser as a function of several parameters, such as the discharge current, composition and pressure in the working mixture, etc. The results of these calculations gave a quantitatively correct description of the experimental dependences, which confirmed the main physical ideas on the operation of the CO_2 laser. However, the quantitative agreement was poor because the experimental and theoretical results differed by an order of magnitude. Moreover, it became clear that the theory could not be developed further without a comprehensive study of the properties and parameters of the active medium of the carbon dioxide laser, which was the discharge plasma in a multicomponent molecular gas mixture. Until recently, relatively little work had been done on this type of discharge.

One of the central problems associated with the specific features of discharges in complex mixtures of molecular gases is the determination of charges in the original components under the action of the discharge. The importance of the influence of these changes on the properties of the active medium follows from the characteristic features of the CO_2 laser: If a high and constant output power is required, a gas mixture has to be supplied continuously to the discharge gap. The first attempts to construct sealed CO_2 lasers (without continuous flow) were either unsuccessful or yielded much lower output powers compared with those available

---

*Detailed reviews of the work published up to 1967 were given in [5, 6].

from continuous-flow devices; moreover, the service life of sealed CO_2 lasers was found to be limited.

It should be remembered that changes in the molecular composition of gases under the action of a discharge may also play a positive role in the population inversion mechanism. For example, it was first reported in [5, 9] that, in principle, the CO molecules formed as a result of dissociation of CO_2 might participate in the population of the upper laser level. However, the quantitative contribution of this process was not clear because of the lack of data on the dissociation of carbon dioxide.

Since this investigation was started (May 1967), only one brief communication on the chemical equilibrium in a sealed CO_2 laser has been published [13]. At the same time, the first results of an investigation of chemical reactions in a sealed system, carried out at the Institute of Petrochemical Synthesis in collaboration with the Lebedev Physics Institute, were published [14]. These first studies showed that in the investigated cases, the original gas mixture suffered considerable changes, mainly because of the dissociation of the CO_2 molecules. However, these results were tentative and incomplete (we shall consider them later in detail). Obviously, it was necessary to carry out a systematic study of the dissociation of carbon dioxide molecules under the discharge conditions of the type employed in CO_2 lasers. This was the task of the investigation reported below.

CHAPTER I

DISSOCIATION OF CO_2 MOLECULES IN A CO_2 LASER GAS-DISCHARGE PLASMA

The aim was to carry out a detailed and systematic experimental study of the processes resulting in changes in the chemical composition of the active medium in a CO_2 laser under the action of electrical discharges, to determine the nature of the influence of these changes on the properties of the active medium, and to review (on this basis) ideas on the physical processes resulting in population inversion.

§ 1. Brief Review of Investigations of
Dissociation of CO_2 in Electrical Discharges

The dissociation of CO_2 molecules in electric discharges has been studied by many workers. However, the results cannot be applied directly to the CO_2 laser because the conditions in such experiments have usually been different from those in electrical-discharge carbon dioxide lasers.

The first quantitative determinations of the degree of dissociation of CO_2, carried out mainly by manometric methods, frequently gave contradictory results (a review of the early work on the dissociation of CO_2 can be found, for example, in Shekhter's book [15]).

More reliable data on the dissociation of CO_2 in discharges were obtained as a result of the development of modern methods for gas analysis, i.e., gas chromatography and mass spectrometry. Analyses carried out under various electrical discharge conditions have demonstrated that the dissociation products of carbon dioxide are carbon monoxide and oxygen. Hardly any ozone forms in low-pressure discharges and carbon monoxide does not dissociate [16]. Therefore, the chemical reactions in a discharge zone can be described by the simple system

$$CO_2 \rightleftarrows CO + O, \tag{1}$$

$$O + O \rightleftarrows O_2. \tag{2}$$

If thermal dissociation takes place, the relationship between the concentrations of the original substances and of their dissociation products can be deduced from the known equilibrium constants [17]. With this point in mind, a comparison is made in [18, 19] of the observed degrees of dissociation $\alpha = [CO]/[CO] + [CO_2]$ with the degrees of dissociation corresponding to equilibrium conditions at the gas temperature in the discharge. It is found that the observed degrees of dissociation α considerably exceed the equilibrium values and hence the dissociation of CO_2 (and oxygen) occurs under the action of electron impact:

$$CO_2 + e \rightarrow CO + O + e, \tag{1a} *$$

$$O_2 + e \rightarrow O + O + e. \tag{2a}$$

The reverse reactions occur as a result of thermal recombination in triple collisions:

$$CO + O + M \rightarrow CO_2 + M, \tag{1b}$$

$$O + O + M \rightarrow O_2 + M. \tag{2b}$$

The value and characteristic times for the establishment of the steady-state degree of dissociation α_∞, i.e., the degree which should be reached in a discharge after an infinitely long time, are governed by the relationship between the dissociation and recombination rates.

In the absence of data on the probabilities of elementary dissociation processes, it is impossible to calculate theoretically the rates and degrees of dissociation of CO_2 in the discharges and one has to turn to experimental studies.

The dependences of the steady-state degrees of dissociation of CO_2 under various conditions in discharges have been investigated. The measurements reported in [20] were carried out in a closed circulation system using discharge currents of 100-600 mA. It was found that when the initial pressure in the system was lowered from 100 to 30 Torr, the steady-state degree of dissociation increased from 48 to 80% for a current of 600 mA. At the lowest pressure of 30 Torr, the degree of dissociation depended weakly on the discharge current, whereas, at higher pressures, it fell quite strongly when the current was reduced. The measurements reported in [21] were carried out at CO_2 pressures of 40, 80, and 160 Torr using a current of 350 mA; they also showed that the steady-state degree of dissociation decreased with rising CO_2 pressure. Similar results were obtained in [18] when the CO_2 pressure was varied within the range 50-300 Torr and the current within the range 100-800 mA. The investigations reported in [19] were carried out using much lower currents. A study of the dependence of α_∞ on the initial CO_2 pressure, carried out in the range from 20 to 600 Torr, demonstrated that this dependence was nonmonotonic. When the pressure was increased, the value of α_∞ first rose, reached a maximum, and then fell, tending to a constant value in the high-pressure range. For a current of 12 mA, this maximum (~ 0.4) was reached at 50 Torr, whereas, for a current of 20 mA, it was attained at 150 Torr.

It follows from these results that extrapolation from the values obtained at high CO_2 pressures (tens and hundreds of torr) to low pressures (~ 1 Torr) is unjustified and may give rise to qualitatively wrong results. An analysis of the published data on the dissociation of CO_2 in discharges shows that most of the investigations have been carried out at high pressures

---

*It should be pointed out that, in addition to the dissociation of CO_2 by electron impact (1a), this molecule may also dissociate because of the heating of the gas in the discharge, and the predominance of one or the other dissociation mechanism depends on the actual conditions during a discharge.

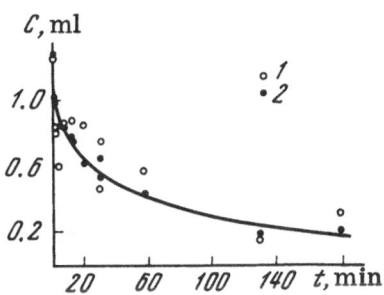

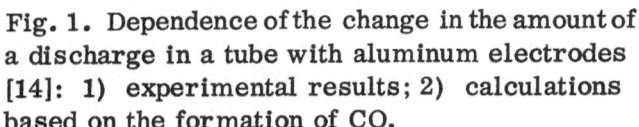

Fig. 1. Dependence of the change in the amount of a discharge in a tube with aluminum electrodes [14]: 1) experimental results; 2) calculations based on the formation of CO.

and, therefore, the results cannot be applied to CO_2 lasers in which partial pressures of carbon dioxide are usually of the order of 1 Torr. A specific feature of discharges in CO_2 lasers is also their geometry: A long and narrow discharge tube is used; moreover, discharges occur not in pure CO_2 but in mixtures of CO_2 with other gases.

It follows that it is desirable to investigate independently the dissociation reactions in discharges employed in carbon dioxide lasers. The first investigation of the changes in the composition of the laser mixture was reported briefly by Macken et al. [13]. This communication gave very little information on the chemical equilibrium in the investigated CO_2 laser. The most important result of the reported investigation, carried out by the mass spectrometric method, was that only the main component of the laser mixture, i.e., carbon dioxide, changed in the discharge.

A more detailed study of a sealed CO_2 laser system was reported in [14]: The gas composition was determined by chromatography. A study was made of changes in the gas composition in a molybdenum glass tube with aluminum electrodes. The experiments were carried out on a CO_2-He mixture (1:10) at a total pressure of 8.2 Torr for a discharge current of 25 mA.

The results reported in [14] are plotted in Fig. 1. It is clear from this figure that the loss of CO_2 was very rapid and there was very little carbon dioxide left after 3 h. Clearly, stimulated emission stopped in a similar or shorter period. It should be noted that, within the limits of the experimental error ($\pm 15\%$ in the case of CO_2 and $\pm 3\%$ in the case of CO), the amount of CO formed in the discharge corresponded to the $CO_2 \rightleftharpoons CO + O$ reaction, i.e., the reaction was balanced in respect of carbon. This demonstrated once again that the $CO \rightarrow C + O$ dissociation, if it occurred at all, was negligible. Another important result of this investigation was that the loss of CO_2 occurred in parallel with the loss of oxygen. Conversely, the addition of oxygen resulted in a temporary reappearance of CO_2 and stimulated emission. After a time, both the CO_2 and oxygen disappeared again. This irreversibility was attributed in [14] to the loss of oxygen by adsorption (or chemisorption) on the electrodes or tube walls. The investigation just described was the only detailed study known to the present author at the moment of starting this investigation.

The aim of the present investigation was to study in detail the dissociation of CO_2 in sealed and continuous-flow (including flow at high rates) laser systems as a function of many discharge parameters such as the current, pressure, and original composition of the gas mixtures, discharge geometry, etc.

In the course, and after completion, of this investigation several papers were published abroad on the chemical composition of discharge plasmas in CO_2 lasers. These papers will be discussed at the end.

§ 2. Method for Determining Gas Composition

Various methods can be used in studies of the molecular composition of a multicomponent gas-discharge plasma. One of the most widely used methods of quantitative analysis, emission spectroscopy, meets with considerable difficulties [22]. The main problem is that the investigated dissociation processes alter the gas pressure and change the concentrations of the components with different ionization potentials. This changes the distribution functions of the electron velocities and densities, i.e., it affects the conditions of excitation of spectral lines. It is practically impossible to allow accurately for changes of this kind. Nevertheless, emission spectroscopy can be used successfully (because of its sensitivity) to establish the presence of specific components [23-25].

Modern well-developed analytic methods for gas mixtures also include mass spectrometry and gas chromatography. These methods are sufficiently sensitive and can be used to determine the quantitative amounts of all the components in a mixture. However, these methods are time-consuming and relatively complex. Moreover, additional difficulties associated with their use appear in studies of steady-state systems. This is because the methods themselves require the taking of samples of gas, i.e., the system can no longer be closed. For example, in order to determine changes in the concentrations of the components as a function of the duration of service life of a sealed laser, it is necessary either to alter many times the geometry of the laser or repeat as many times the experiment right from the beginning [14, 25-27]; this reduces the reproducibility. Classical methods of analytic chemistry involving quantitative reactions suffer from the same disadvantages and are even more time-consuming. We therefore adopted a different method for monitoring the main component, i.e., the CO_2 gas, in which we measured the absorption of infrared radiation by carbon dioxide molecules in their ground state.

Infrared Gas Analyzer

We measured the concentrations by comparing the intensities of two infrared beams of different wavelengths after passing through the investigated gas [28]. The wavelength of one of the beams was selected so as to avoid any absorption bands of the investigated gases, whereas the wavelength of the other beam was deliberately tuned to the center of the absorption band of the investigated gas.

The transmitted infrared beams were directed in turn to a detector. The electrical signals produced by this detector were amplified and switched so as to balance an electric circuit. A reduction in the intensity of one of the beams unbalanced the circuit and the unbalance was a measure of the change in the investigated concentration.

The apparatus is shown schematically in Fig. 2. A source of light 1 was a calibrated Nichrome helix wound on a ceramic tube. The influence of convection in the gas was avoided by

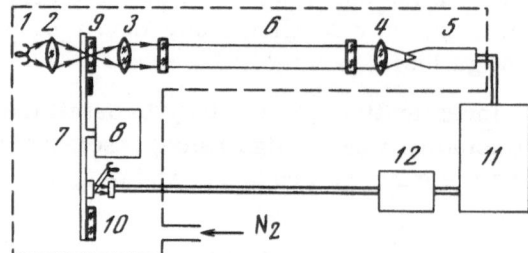

Fig. 2. Schematic diagram of an infrared gas analyzer.

covering the source with a mica screen. Lithium fluoride lenses 2-4 focused the light from this source onto a BSG-2 bolometer 5. A cell 6, containing the investigated gas, was placed between the lenses 3 and 4; the light was passed through a cell in the form of a parallel beam. The cell ends were closed by calcium fluoride (fluorite) windows. The light was modulated by a disk 7 rotated by a synchronous motor 8 (we took a motor from a PPCh-1 unit employed for daily checks of clocks). The disk had two apertures covered by narrow-band interference filters 9 and 10. The transmission band of one of the filters had a maximum at 3.97 μ; the second filter was selected depending on the concentration of the investigated gas: In the measurements of the CO_2 concentration, the filter employed had a transmission maximum at 4.3 μ, whereas, in the case of CO, the maximum was at 4.62 μ, which corresponded to the strongest absorption bands.

The filters were made at the State Optical Institute; their maximum transmission was about 60% and the half-width of the transmission band was about 0.07 μ. All three filters had transmission bands in adjacent spectral ranges (this reduced the influence of the instabilities of the light source). Moreover, all the filters were used in the region of the maximum emissivity of the source.

Electrical signals produced by the bolometer were applied first to a V6-4 standard amplifier and then to an electronic circuit 11 in which the signals corresponding to the two beams were compared and subtracted. This circuit was operated by a relay whose winding was supplied with Π-shaped pulses generated by a photoresistor 10 and passed through an auxiliary amplifier 12. This photoresistor was illuminated with modulated light from a small incandescent lamp by reflection from a black-and-white circle bonded to the rotating disk. The recording unit was an ÉPPV-28 automatic plotter with a large input resistance.

All the parts within the dashed line in Fig. 2 were enclosed in a jacket. Compressed nitrogen was blown continuously through this jacket during measurements of the CO_2 concentrations and this was done in order to avoid the influence of atmospheric CO_2.

The system was calibrated before measurements. This was done by admitting a gas at different pressures into the cell 6 (the pressure was measured with an oil manometer). It was found that the readings of the recording unit were a linear function to the CO_2 pressure (in the 0-10 Torr range). Since the relative precision of the pressure measurements was fairly good at high pressures, the linearity of the calibration made it possible to estimate the precision (associated only with the calibration process), which was $\pm 0.5\%$ in respect to the CO_2 concentrations and $\pm 5\%$ in respect to CO. A special check showed that in the determination of the concentration of one of the components the influence of the absorption bands of the other components was completely eliminated by the use of the interference filters with narrow transmission bands.

Measurement Method

The system described above could be used to determine the CO_2 and CO concentrations during a discharge in the cell 6. This was possible because the electronic circuit effectively performed two-channel synchronous detection of the signal and, under dc conditions, the discharge signal could not be amplified. However, basic difficulties appeared when the concentrations were measured directly in the discharge. These difficulties were primarily due to the special nature of the active medium of the CO_2 laser.

The main difficulty arose from intensive excitation of the vibrational levels of molecules in the CO_2 discharge plasma. For example, under near-optimal conditions, there were usually ~10% of the total number of molecules at the upper laser level 00^01 whereas, at the 01^10 level, there were about 15% of these molecules when the discharge temperature was 500°K. An accurate indirect allowance for the presence of molecules in vibrational levels was very difficult to make and it was one of the main problems encountered in the identification of the population inversion mechanism of the CO_2 laser (a more detailed discussion of the populations of the vibrational levels is given in Chaps. II and III). The absorption method employed in the present study allowed us to determine directly only the concentration of the molecules in the ground state.

In view of this difficulty, we did not measure the concentrations directly in the discharge. For example, in the case of continuous-flow systems, the concentrations of the components were determined at the exit from the discharge, whereas, in sealed systems, the discharge was temporarily switched off during the measurements. When this measurement method was adopted, it was important to determine to what extent the concentrations of the CO and CO_2 molecules measured with the discharge switched off (or outside the discharge zone) corresponded to the concentrations during the discharge.

We could show that this correspondence was quite accurate, i.e., that the concentration of the CO and CO_2 molecules determined when the discharge was switched off differed only slightly from the concentrations during the discharge. Possible factors which could alter the chemical composition of the gas when the discharge was switched off included the thermal recombination of the products of the dissociation of CO_2, reactions with short-lived, highly reactive radicals, and reactions due to long-lived particles in the discharge. We shall now estimate to what extent these processes could alter the concentrations of the CO and CO_2 molecules.

When the discharge was switched off, some of the CO molecules recombined with atomic oxygen but the number of such molecules was small. This was due to the fact that the rate constant of the $O + O + M \xrightarrow{K_1} O_2 + M$ reaction was considerably greater than the rate constant of $O + CO + M \xrightarrow{K_2} CO_2 + M$. The values of K_1 and K_2 ($cm^6 \cdot mole^{-2} \cdot sec^{-1}$) taken from different sources were as follows:

K_1		K_2		
$8.4 \cdot 10^{-33}$ (M=CO_2) [29]	$8 \cdot 10^{-35}$	at	$T = 428°$ K	[32]
$3.9 \cdot 10^{-33}$ (M=N_2O) [29]	$8,5 \cdot 10^{-35}$	at	$T = 293°$ K	[33] *
$2.8 \cdot 10^{-33}$ (M=N_2) [29]	$1.2 \cdot 10^{-35}$	at	$T = 300°$ K	[34] †
$8.4 \cdot 10^{-34}$ (M=He) [29]	$6 \cdot 10^{-35}$	at	$T = 400°$ K	[34]
$1.1 \cdot 10^{-33}$ (M=He) [30]				
$3.1 \cdot 10^{-33}$ (M=N_2) [30]				
$2.7 \cdot 10^{-33}$ [31]				

It is concluded in [29, 31] that K_1 depends weakly on temperature.

It follows from the above table that when a discharge is switched off, oxygen atoms recombine preferentially with one another and not with carbon monoxide molecules. Moreover, since $C_{CO} > C_O$ (because at the moment when a discharge is switched off a considerable proportion of oxygen is in the molecular state), it becomes clear that the concentration of the CO (and, consequently, of CO_2) molecules can only change slightly after the end of a discharge. (According to approximate estimates obtained by the present author, the partial recombination of CO and O gives rise to a difference not exceeding 2–3% between the concentrations of CO_2 during a discharge and after switching it off.)

The problem of a possible influence of active particles (radicals and excited species) on chemical reactions after the end of a discharge is very difficult and requires special study. We shall give some qualitative reasons for assuming that there are only slight changes in the concentrations of the CO and CO_2 molecules due to participation of such active particles in the reactions.

---

*M = O_2; the values for M = N_2, He, Ar, and Kr are even smaller.

† Temperature dependence of this constant is $K_2(T) = 2.2 \cdot 10^{15} \exp(-3700/RT)$ $cm^6 \cdot mole^{-2} \cdot sec^{-1}$.

The presence of any hypothetic reactive radicals cannot alter significantly the concentration of the dissociation products after the end of a discharge. The high reactivity of such radicals ensures that their concentration is low and, whereas in a discharge zone their presence may affect chemical reactions decisively (because the steady-state concentration of such radicals is maintained by the discharge energy), they should disappear quite rapidly after the end of a discharge (the concentration of OH radicals in a discharge in the presence of water vapor or hydrogen is discussed in § 4 in the present chapter).

The absence of thermodynamic equilibrium, which is characteristic of electric discharges, ensures efficient excitation of electronic states of atoms and molecules which have sufficient energy for chemical reactions to take place. Obviously, changes in the chemical composition of a gas after the end of a discharge can only be due to particles excited to metastable states. For example, it is reported in [35] that when CO_2 is mixed with active nitrogen, weak dissociation of carbon dioxide is observed because of the long-lived nitrogen state $A^3\Sigma_u^+$.

An analysis of the excitation conditions shows that, in the case of laser mixtures containing nitrogen, the $A^3\Sigma_u^+$ state is populated more than any other metastable electronic state but this population does not exceed 10^{12} cm^{-3} (see Chap. III). This is four orders of magnitude lower than the concentration of the CO_2 molecules under conditions typical of CO_2 laser discharges. Since the energy of the $A^3\Sigma_u^+$ state is about 6 eV, which is comparable with the dissociation energy of CO_2 (~ 5.5 eV), changes in the concentration of CO_2 after the end of a discharge due to such excited nitrogen should not exceed the concentration of the excited $N_2(A^3\Sigma_u^+)$ molecules, i.e., the relative change should be $\sim 10^{-4}$.

Thus, our analysis shows that the method of measuring the CO_2 and CO concentrations outside a discharge is justified and we may assume that the concentrations of CO and CO_2 in a discharge zone should remain practically the same after switching off the discharge.*

§ 3. Dissociation of CO_2 in a Continuous-Flow

Laser System

Experimental Method

We investigated a continuous-flow system with water-cooled molybdenum glass tubes of different diameters. We used hollow cylindrical molybdenum electrodes placed in side branches. The upper edges of the electrodes were located 45 mm from the tube axis. In all the tubes, the discharge gap was 600 mm long. With the exception of the narrowest tube (internal diameter 3 mm), all the tubes were closed with NaCl windows oriented at the Brewster angle. They were placed inside a resonator formed by two external gold mirrors whose radii of curvature were 15 m. An aperture for the extraction of laser radiation was made in one of these mirrors.

The investigated tubes were connected in series with a gas-analyzer cell so that the mixture which crossed the discharge gap reached the analyzer cell after passing through a connecting tube of ~ 15 mm diameter and about 30 cm long.

---

* The validity of our method is limited to relatively low gas temperatures in the discharge. It is clear from the temperature dependences of the rate constants K_1 and K_2 that when the temperature is increased, the value of K_2 rises rapidly and the difference between the rates of the CO + O and O + O recombination processes decreases. At $T \gtrsim 1000°K$, these rates become comparable. Under conditions typical of CO_2 lasers, the gas temperatures are much lower.

The gas pressure was measured with an oil manometer on entry of the gases into the discharge tube and the error in these measurements was up to ±0.035 Torr. Under continuous-flow conditions, there was a pressure drop along the discharge tube. This drop was measured by connecting to both ends of a tube the two arms of a differential oil manometer which recorded the difference between the pressures at entry and exit from the tube. Measurements showed that in the investigated range of gas flow velocities the pressure gradients in tubes of 34 and 22 mm diameter were negligible. In an 8-mm diameter tube, the pressure drop did not exceed 5%. However, in the case of the narrowest tube (3-mm internal diameter), the pressure drop was very large (~80% at the maximum flow velocities). When this tube was used, we reduced the flow velocity so that the pressure drop did not exceed 10% of the pressures of gases at the entry to the tube.

The flow of the gas and its linear velocity were measured as described in Appendix I.

The greatest errors arose in measurements of the flow velocity and were primarily due to inaccuracies in the values of the gas temperature; these errors were ±10%. This resulted in an uncertainty in the CO_2 concentration not exceeding ±3%. The discharges were supplied with a direct current of 0–40 mA and the applied voltage was up to 12 kV. The depth of modulation of the voltage did not exceed 1-2% and its frequency was 100 Hz.

Experimental Results and Discussion

Let us now consider in detail the results obtained for a tube of 22 mm diameter and 60 cm length.

Figures 3a and 4a give the CO_2 concentrations as a function of the rate of flow, total pressure in the mixture, and discharge current for $CO_2 - He$ and $CO_2 - N_2 - He$ mixtures. The concentration of CO_2 in the absence of a discharge was taken as 100% so that $C(CO_2) = (1 - \alpha) \cdot 100\%$, where α is the degree of dissociation of the CO_2.

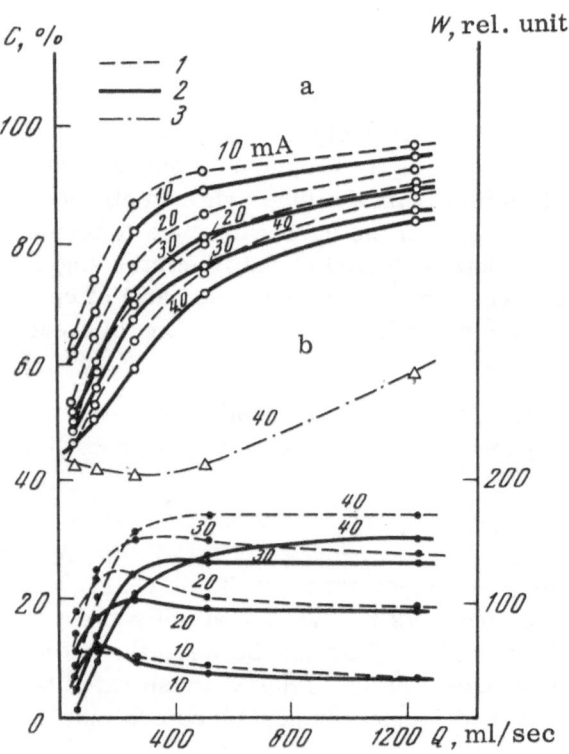

Fig. 3. Dependences of the CO_2 concentration (a) and laser output power (b) on the rate of flow of the gas plotted for different pressures p = 8.3 (1), 6.9 (2), and 3.6 Torr (3) and different discharge currents (values given alongside the curves) in a $CO_2 - He$ (1 : 9) mixture.

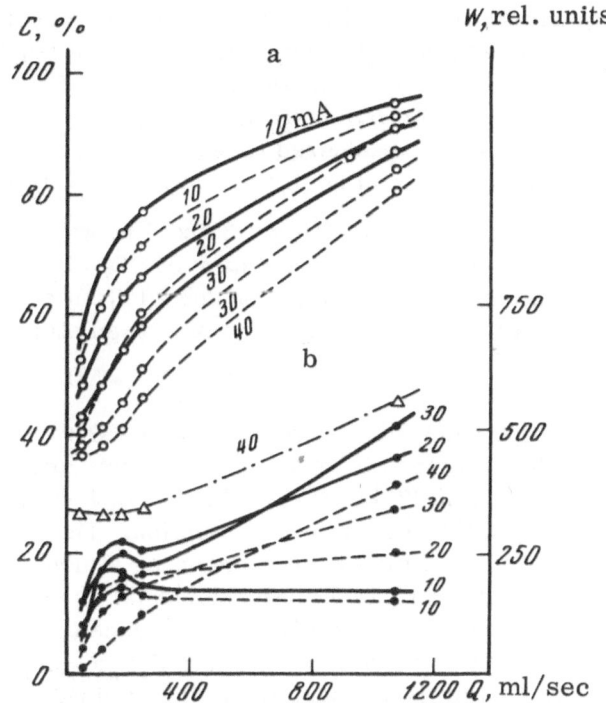

Fig. 4. Dependences of the CO_2 concentration (a) and laser output power (b) on the rate of flow of the gas plotted for different pressures and discharge currents in a CO_2-N_2-He mixture (1 : 3 : 6).

It is clear from these figures that the CO_2 concentration decreases when the gas flow rate or the total pressure decreases and when the discharge current is increased. Figure 5 shows the dependence of the CO_2 concentration in a binary CO_2-He mixture on the discharge current for two values of the rate of flow and pressure. It is clear from Fig. 5 that the dependence is steeper in the low-current range. This is manifested particularly clearly when the pressure is reduced and especially when the rate of flow is lowered. This can be explained by the fact that the degree of dissociation increases with decreasing pressure and rate of flow (Fig. 3) so that the CO_2 dissociation reaction

$$CO_2 \rightleftarrows CO + O$$

approaches equilibrium even at low currents.

In studies of the processes occurring in a discharge, we must allow for the role of the walls and electrodes. Apart from the oxidation of CO, which is the reverse of the dissociation of CO_2, we may encounter oxidation of CO by oxygen desorbed from the wall and electrodes and, possibly, the reaction of oxygen with carbon which may be present in small amounts on the electrodes and walls. When the gas is flowing slowly, these reactions may even increase the concentration of carbon dioxide when the rate is reduced still further because free oxygen then spends a longer time in the discharge zone (i.e., the

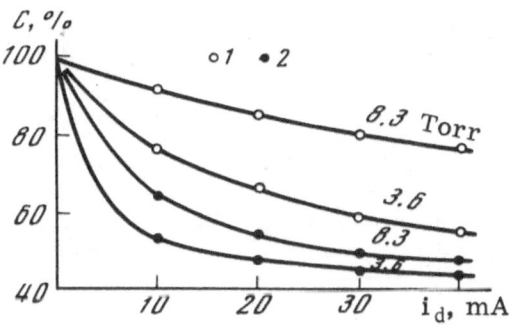

Fig. 5. Dependences of the CO_2 concentration in a CO_2-He mixture on the discharge current at different pressures (given alongside the curves) and for different flow velocities: 1) Q = 500 ml/sec; 2) 55 ml/sec. The tube diameter was 22 mm and its length was 60 cm.

balance in respect of carbon is no longer observed). The contribution of such reactions may be considerable at low pressures and high currents. This accounts for the dependences of carbon dioxide concentration on the rate of flow at a pressure of p = 3.6. Torr for a discharge current i_d = 40 mA. Conversely, at high pressures and flow rates, we may assume that the carbon balance is adhered to and the relative content of the CO molecules is C(CO) = 100% − C(CO$_2$), i.e., we can ignore adsorption and desorption.

A comparison of Figs. 3 and 4 shows that at low flow rates of CO$_2$−N$_2$−He ternary mixtures, the degree of dissociation of CO$_2$ is higher than in the binary CO$_2$−He case. This may be explained by assuming that the addition of nitrogen produces NO$_2$ molecules, which act as catalysts of the recombination of oxygen atoms [36]:

$$NO + O \rightarrow NO_2, \quad NO_2 + O \rightarrow NO + O_2,$$

shifting the equilibrium of the CO$_2 \rightleftarrows$ CO + O reaction to the right. However, it should be noted that NO is formed in very small amounts and can be detected only because of its visible emission spectrum [37]. On the other hand, it is shown in [36] that even small amounts of NO$_2$ have a strong influence on the dissociation. Thus, the addition of NO$_2$ in amounts below 0.5% of the total gas in the mixture can suppress almost completely the oxidation of CO. The presence of NO in the investigated mixtures cannot be detected by mass spectrometry or chromatography [17, 37, 38]. However, there is evidence that N$_2$O is formed in discharges. For example, it is reported in [39] that in addition to the CO$_2$-stimulated emission lines, there were also lines due to CO$_2$ in those cases when the original mixture contained nitrogen. An interesting result reported in [39] was that the stimulated emission was strongest for the rotational lines with low values of the rotational quantum numbers J ∼ 2-3, although no deliberate selection was made of the rotational lines in the resonator. On the other hand, it was reported in [40] that an investigation of the stimulated emission from N$_2$O molecules present originally in a mixture established that the emission maximum corresponded to rotational lines with J ∼ 20, which agreed with the maximum population at the gas temperature in the discharge. Therefore, the stimulated emission maximum corresponding to transitions with low values of J, reported for N$_2$O formed chemically in a discharge, could best be explained by the fact that the molecules were formed in the states with low values of J in such discharges. The thermalization of the rotational levels did not occur because of their very short lifetimes.

We also attempted to determine the concentration of the CO$_2$ molecules using the ∼ 4.5 μ absorption band. Our measurements indicated that if N$_2$O was present, its amount was less than $5 \cdot 10^{-2}$ Torr (the total pressure in the gas mixture was ∼10 Torr).

A clear idea on the rate of change in the CO$_2$ concentration can be gained from Fig. 6, which shows the dependences of the changes in the CO$_2$ concentration in CO$_2$−He (1:9) and CO$_2$−N$_2$−He (1:3:6) mixtures at a total pressure of p = 6.9 Torr on the time spent by the mixture in the discharge tube. These curves were plotted on the basis of the results given in

Fig. 6. Dependences of the CO$_2$ concentration in CO$_2$−He (continuous curves) and CO$_2$−N$_2$−He (dashed curves) mixtures on the time spent by the mixture in the discharge, plotted for i_d = 20 (1), 30 (2), and 40 mA (3).

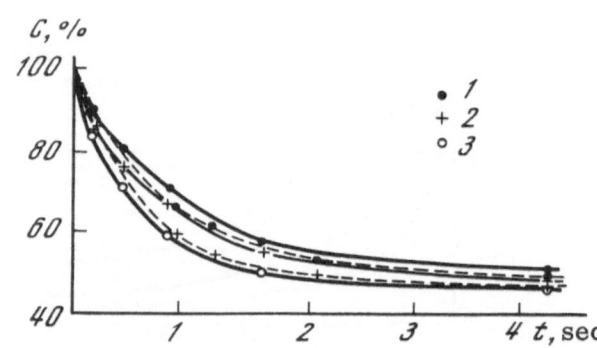

Figs. 3 and 4 (the volume of the tube was 230 cm^3). The $CO_2 \rightleftarrows CO + O$ equilibrium was reached after several seconds.

We studied the dissociation of CO_2 in the mixtures mentioned above and in pure CO_2 and $CO_2 - N_2$ (1 : 3). In spite of the fact that in the case of CO_2 and $CO_2 - N_2$ the rates of flow of the discharge could not be determined sufficiently accurately because of lack of experimental data on the gas temperatures (see Appendix I), we concluded that for identical partial pressures in CO_2 the strongest dissociation of carbon dioxide occurred in $CO_2 - N_2$ and the weakest in $CO_2 - He$. The dissociation in $CO_2 - N_2 - He$ was less than in pure CO_2, i.e., the addition of He reduced the degree of dissociation of CO_2.

Figures 3b and 4b give the dependences of the stimulated emission power (in relative units) on the rate of flow of the gas, pressure, and discharge current in $CO_2 - He$ and $CO_2 - N_2 - He$ mixtures. We can see that the output power depends strongly on all these parameters. It is worth noting that, in some cases, there are maxima at low flow rates. This may easily be explained bearing in mind the population inversion mechanism suggested in [5, 9, 11]. According to the hypothesis put forward in these papers, a strong pumping of the upper laser level 00^01 of the CO_2 molecule is due to a resonant transfer of energy from vibrationally excited CO molecules to CO_2 molecules (this point will be discussed in greater detail in Chaps. II and III).

Comparison of the dependences of the changes in the CO_2 concentration and output power in Fig. 3 shows that all the maxima for the $CO_2 - He$ mixture agree well with the ratio $C(CO_2) : C(CO) = 7 : 3$ if we assume that $C(CO) = 100\% - C(CO_2)$. This also explains the results reported in [41]; in this investigation, the heating of a platinum wire inside a discharge tube reduced considerably and sometimes even suppressed the stimulated emission if the working mixture did not contain N_2. This was due to the fact that platinum catalyzed the oxidation of CO to CO_2 so that the population inversion mechanism was quenched.

In the case of the ternary $CO_2 - N_2 - He$ mixture, the population inversion was mainly due to the vibrationally excited N_2 molecules. However, it is clear from Fig. 4b that, as in the binary mixture case, the stimulated emission has maxima at low rates of gas flow and these maxima are much more pronounced than in the binary case.

A reduction in the degree of dissociation in a continuous-flow system with increasing rate of flow of the gas and an increase with increasing discharge current were reported in [42]. It was established there that an increase in the rate of flow of the gas resulted in an increase of the current at which the output power had its maximum value. This was due to the fact that under optimal (from the point of view of population inversion) conditions there should be a certain amount of CO in the mixture.

Smith [38] obtained similar results. He used a discharge tube with an internal diameter of 25 mm and length of 2 m; it was fitted with aluminum electrodes which were concentric with the tube axis. The gas composition was measured by mass spectrometry at the exit from the discharge zone. On the whole, the conclusions reached by Smith agreed with our results, i.e., the degree of dissociation increased when the flow rate and pressure were reduced and when

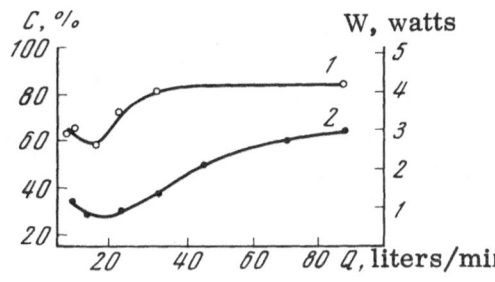

Fig. 7. Dependences of the laser output power (1) and CO_2 concentration (2) on the rate of flow of a $CO_2 - N_2 - He$ (6 : 12 : 8) mixture at p = 9 Torr, i_d = 70 mA [38].

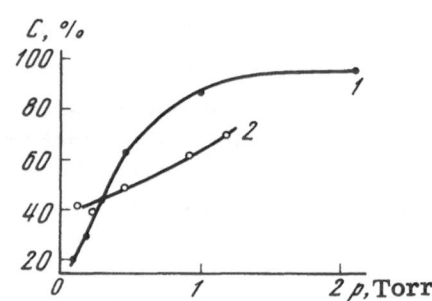

Fig. 8. Dependences of the CO_2 concentration on the partial pressure of carbon dioxide in a mixture flowing at 90 liters/min in a tube 25 mm in diameter and 2 m long: 1) pure CO_2, $i_d = 37$ mA; 2) $CO_2 - N_2 - He$ (6 : 12 : 8), $i_d = 70$ mA [39].

the discharge current was increased. Smith also found some singularities in the behavior of the dependences of the CO_2 concentration and output power on the gas flow at low rates of this flow. It is clear from Fig. 7 that the latter effect was manifested in [38] more clearly than in our results (compare with Figs. 3 and 4). This could be explained by the fact that Smith [38] used aluminum electrodes which exhibited a considerable sputtering in discharges, i.e., the active surfaces of these electrodes were larger. Moreover, whereas in our case the molybdenum electrodes were placed inside the tubes, the gas in the tube used by Smith flowed directly through the electrodes. Smith attributed the observed features to the fact that at low rates of flow the traces of water vapor or hydrogen played an important role (the role of water and hydrogen is discussed in greater detail in § 3 in the present chapter).

Other results obtained by Smith [38] are plotted in Fig. 8. This figure gives the dependence of the rate of decomposition of CO_2 on the pressure in pure CO_2 and in a $CO_2 - N_2 - He$ mixture. A comparison of the degree of dissociation in mixtures of different compositions led Smith [38] to the conclusion that the addition of nitrogen accelerated the dissociation of CO_2. He also concluded that the addition of helium had the same effect. This contradicted our results. However, the cause of this contradiction could easily be explained. The point was that the degree of dissociation in different mixtures could be compared only under identical conditions, including the same linear velocities of the gas flow in the discharge so that the gas spent the same time in the discharge zone. In the work of Smith [38], the velocity was measured outside the discharge tube and there was no indication that this was the true velocity in the discharge because the latter depended on the gas temperature (see Appendix I). The addition of helium was known to reduce considerably the gas temperature [43] and, therefore, the true gas velocity; this could result in an apparent increase in the rate of dissociation if the velocity were measured outside the discharge zone.

Measurements similar to those described above were also carried out in tubes of different diameters. The general relationships (the fall in the CO_2 concentration as a result of a reduction in the rate of flow or pressure in the discharge zone and as a result of increase in the discharge current) were the same for all the tubes. It should be noted that when the tube diameter was reduced, the rate of dissociation of CO_2 increased considerably. For example,

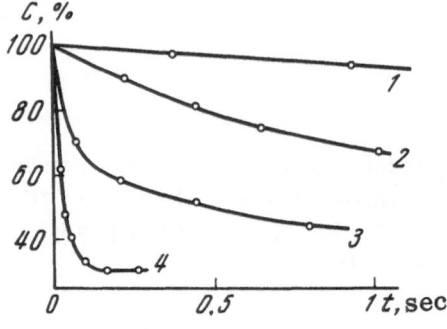

Fig. 9. Dependences of the CO_2 concentration on the time spent by a $CO_2 - N_2 - He$ (2 : 1 : 8) mixture in discharges occurring in tubes of different diameters (mm): 1) 34; 2) 22; 3) 8; 4) 3. The pressure in the tube was p = 5.5 Torr and the current was $i_d = 10$ mA.

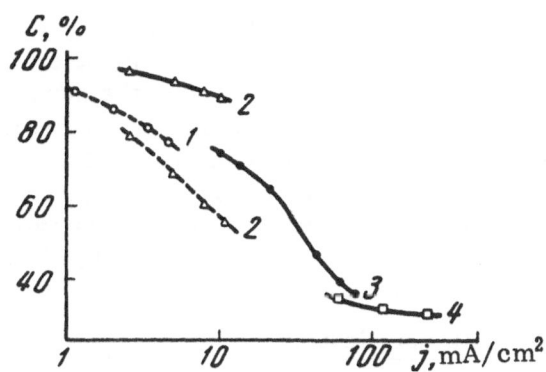

Fig. 10. Dependences of the CO_2 concentration in a $CO_2 - N_2 - He$ (2 : 1 : 8) mixture on the current density in tubes of different diameters. The notation is the same as in Fig. 9. The continuous curves correspond to p = 5.5 Torr, v = 3 m/sec; the dashed curves correspond to p = 6.9 Torr, v = 0.4 m/sec.

in the case of a tube of 34 mm diameter, the dissociation—oxidation reaction (in a $CO_2 - N_2 - He$ mixture of 2 : 1 : 8 composition at a total pressure of p = 5.5 Torr and for a discharge current i_d = 10 mA) reached an effective equilibrium after 15 sec in the discharge, whereas, under the same conditions in an 8-mm-diameter tube, this time decreased to 0.5 sec and in a 3-mm-diameter tube it was only 0.1 sec. Figure 9 shows the dependences of the change in the concentration of carbon dioxide on the time spent by the gas in the discharge zone in tubes of different diameters.

Figure 10 gives the dependences of the CO_2 concentration on the current density in tubes of different diameters. It is clear from this figure that the curves obtained for the same pressures in the mixture and for the same linear velocities of the gas flow did not coincide at the points where the gas densities were equal for tubes of different diameters. The degree of dissociation of CO_2 was greater in narrow tubes. This was due to an increase in the average energy of the electrons in narrower tubes [49].

Role of Different Discharge Zones

In considering the mechanism of stimulated emission from CO_2, it is usual to restrict the analysis to the processes occurring in the positive column of a glow discharge. The results reported above were obtained using electrodes located in side tubes because this reduced somewhat the role of the other discharge zones (those outside the positive column) in continuous-flow systems. The following experiment was carried out to determine the role played by these zones in the dissociation process. Hollow cylindrical molybdenum electrodes were placed on the axis of a long discharge tube and one of these electrodes was moved along the tube. The reduction in the discharge length took place at the expense of the positive column

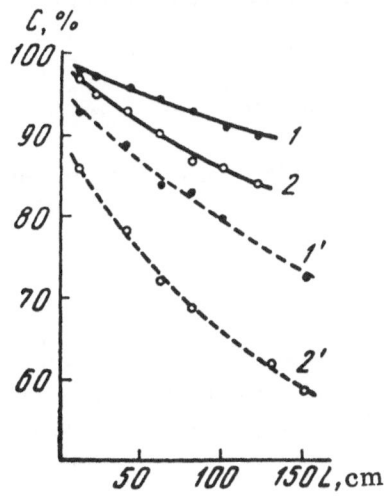

Fig. 11. Dependences of the CO_2 concentration on the length of the discharge gap in a $CO_2 - N_2 - He$ (2 : 1 : 8) mixture flowing through a tube of 25 mm diameter: 1) i_d = 10 mA, p = 6.9 Torr, v = 1.4 m/sec; 1') i_d = 10 mA, p = 3.4 Torr, v = 1.2 m/sec; 2) i_d = 20 mA, p = 6.9 Torr, v = 1.4 m/sec; 2') i_d = 20 mA, p = 3.4 Torr, v = 1.2 m/sec.

[44] so that it was possible to determine the efficiency of the dissociation of CO_2 in the positive column of the discharge and in the other zones. Figure 11 gives the dependences of the CO_2 concentration on the position of the mobile cathode (the length of the discharge given as the abscissa was measured from the cathode). Extrapolation of the dependences to L = 0 showed that the curves did not converge at 100%. Hence, the dissociation in a glow discharge occurred somewhat more rapidly outside the positive column (probably in the cathode dark space). However, it is clear from Fig. 11 that for typical dimensions of the discharge gap used in CO_2 lasers (hundreds of centimeters), the main contribution to the dissociation was made by the positive column.

Dissociation Kinetics

The measured values of the degree of dissociation in the plasma of gas discharges in CO_2 lasers cannot be attributed to thermal processes occurring in a hot discharge gas because the gas temperature is relatively low.

The degree of dissociation of a gas heated under thermodynamic equilibrium conditions can be calculated quite readily because, for each of the reactions, the concentrations of the reacting components are linked by the equilibrium constants.

For example, in the case of the reactions

$$CO_2 \rightleftharpoons CO \tfrac{1}{2} O_2, \quad K_1 = \frac{[CO][O_2]^{1/2}}{[CO_2]}, \tag{1.1}$$

$$\tfrac{1}{2} O_2 \rightleftharpoons O + O, \quad K_2 = \frac{[O]}{[O_2]^{1/2}}, \tag{1.2}$$

the equilibrium constants K_1 and K_2 are known for a wide range of temperatures [17].

In numerical calculations of the concentrations of the separate components, we can use, for example, the method suggested in [45]. We start with the self-evident relationship

$$[O] + [CO_2] + [CO] + [O_2] = p. \tag{1.3}$$

Here, the brackets denote the partial pressures of each of the components and p is the total pressure. We shall also assume that the carbon and oxygen balance conditions are obeyed, i.e.,

$$\frac{n_C}{n_O} = \frac{1}{2} = \frac{[CO_2] + [CO]}{2[CO_2] + [CO] + 2[O_2] + [O]}. \tag{1.4}$$

If the independent variable is the concentration of oxygen atoms [O], we find that Eqs. (1.1)-(1.4) lead to the following third-order equation

$$\frac{2}{K_2^2}[O]^3 + \left(1 + 3\frac{K_1}{K_2}\right)[O]^2 + 2K_1K_2[O] - K_1K_2p = 0. \tag{1.5}$$

If we know the concentration of atomic oxygen, we can readily find the ratio of the amounts of the CO and CO_2 molecules:

$$\frac{[CO]}{[CO_2]} = \frac{K_1 K_2}{[O]}. \tag{1.6}$$

Equation (1.5) was solved for different temperatures and initial CO_2 pressures. Table 1 gives the results of calculations of $[CO]/[CO_2] = \alpha_\infty/(1 - \alpha_\infty)$, where α_∞ is the steady-state degree of dissociation (the numbers in parentheses in Table 1 give the order of magnitude of a given ratio).

TABLE 1. Values of [CO]/[CO$_2$] under Thermodynamic Equilibrium Conditions

$T°$, K	p, Torr							
	0,1	0,2	0,5	1	2	4	5	10
3000	5.57 (2)	2.80 (2)	1.15 (2)	5.98 (1)	3.20 (1)	1.78 (1)	1.49 (1)	8.84 (0)
2800	1.28 (2)	6.64 (1)	2.89 (1)	1.61 (1)	9.46 (0)	5.84 (0)	5.05 (0)	3.31 (0)
2600	2.52 (1)	1.42 (1)	7.15 (0)	4.52 (0)	2.98 (0)	2.04 (0)	1.82 (0)	1.29 (0)
2400	5.22 (0)	3.38 (0)	2.02 (0)	1.42 (0)	1.02 (0)	7.52 (—1)	6.83 (—1)	5.12 (—1)
2200	1.30 (0)	9.36 (—1)	6.27 (—1)	4.71 (—1)	3.57 (—1)	2.74 (—1)	2.52 (—1)	1.95 (—1)
2000	3.52 (—1)	2.70 (—1)	1.92 (—1)	1.49 (—1)	1.16 (—1)	9.12 (—2)	8.44 (—2)	6.63 (—2)
1800	9.03 (—2)	7.10 (—2)	5.17 (—2)	4.08 (—2)	3.22 (—2)	2.55 (—2)	2.36 (—2)	1.87 (—2)
1600	1.83 (—2)	1.45 (—2)	1.07 (—2)	8.46 (—3)	6.70 (—3)	5 32 (—3)	4,94 (—3)	3.92 (—3)

It is clear from Table 1 that the ratio [CO]/[CO$_2$] decreases rapidly when the gas temperature is lowered or when the initial pressure of the carbon dioxide is increased. No calculations are reported for temperatures below 1600°K because they give very small values of [CO]/[CO$_2$]. In principle, we can find these values because the solutions of Eq. (1.5) at T < 1600°K can be described quite accurately by just two terms. Thus, in this range of temperatures, we have

$$\frac{[CO]}{[CO_2]} = \sqrt[3]{\frac{2K_1^2}{p}}.$$

Calculations indicate that the observed dissociation in a discharge ([CO]/[CO$_2$] ~ 1) cannot be described as being in thermodynamic equilibrium.

During the earlier stages of the dissociation process, when the reactions tending to oxidize CO to CO$_2$ do not yet make a significant contribution, we can confine our attention to the dissociation reaction and use

$$-dN_{CO_2} = kN_{CO_2}dt, \tag{1.7}$$

where k is the dissociation rate constant. Therefore, plotting the dependence of the term ln [N$_{CO_2}$(t)/N$_{CO_2}$(0)] on the duration of a discharge, we should obtain a straight line for short times spent by the gas in the discharge zone. This is indeed true (Fig. 12). The values of k can be determined from the slope of the initial part k = tan α. If we assume that the main process which initiates the dissociation is (in our case) electron impact, the value of k can be represented by k = n$_e$ $\langle v_e \sigma_d \rangle$. Here, n$_e$ is the electron density, v$_e$ the electron velocity, and σ_d is the dissociation cross section for electron impact. Averaging can be carried out over a real distribution of the electron velocities in a discharge.

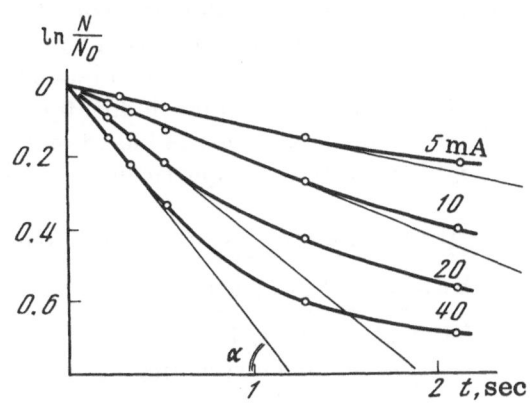

Fig. 12. Dependences of the CO$_2$ concentration on the duration of a discharge in a CO$_2$−N$_2$−He (2 : 1 : 8) mixture, plotted for different currents. The diameter of the tube was 22 mm and the pressure was p = 5.5 Torr.

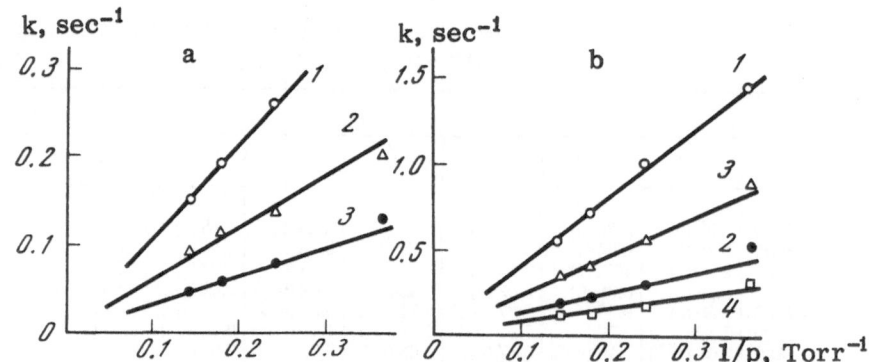

Fig. 13. Dependences of the dissociation rate constant k on the pressure in a CO_2-N_2-He (2 : 1 : 8) mixture in tubes of 34 (a) and 22 mm (b) diameter, plotted for different currents i_d (mA): 1) 40; 2) 20; 3) 10; 4) 5.

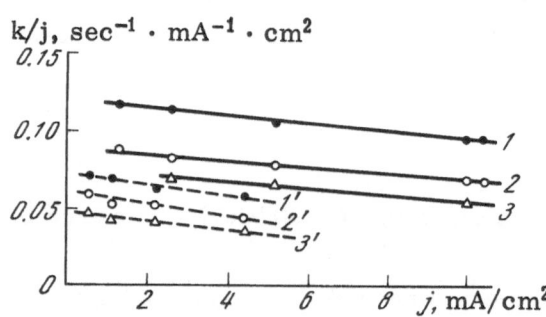

Fig. 14. Dependences of the rate of dissociation, reduced to unit current density, on the current density plotted for different pressures in a CO_2-N_2-He (2 : 1 : 8) mixture in tubes of 22 mm diameter (continuous curves) and 34 mm diameter (dashed curves): 1, 1') p = 4.1 Torr; 2, 2') 5.5 Torr; 3, 3') 6.9 Torr.

The values of k obtained from such kinetic curves are plotted in Fig. 13 as dependences on the pressure in the gas mixture and on the discharge current (regarded as a parameter) in tubes of 34 and 22 mm diameter. It follows from [46] that the electron density is a linear function of the discharge current density so that if the electron velocity distribution remains unchanged, we may expect the value of kj^{-1} to be independent of j. In fact, it is clear from Fig. 14 that the dependence obtained for a constant pressure in tubes of these diameters is very weak. The value of kj^{-1} changes considerably when the gas pressure and tube diameter are altered. These changes are due to a change in the electron velocity distribution which is small when the current density is altered and larger when the pressure and tube diameter are varied. This follows from the experimental results reported in [46]. Unfortunately, there are as yet no detailed data on the nature of these changes. Therefore, we can estimate only approximately the average energy of the electrons in a discharge, which should lie within the range $\bar{\varepsilon}$ = 2-3 eV. We can estimate the cross section σ_d, which may be useful in some approximate calculations. In all the cases under discussion here, the value of σ_d should be within the range $(1-5) \cdot 10^{-18}$ cm^2. The electron density is $n_e = 0.5 \cdot 10^9 j$ cm^{-3} if j is measured in the usual units (mA/cm^2). This expression is based on the results reported in [46] obtained under conditions similar to those in our experiments.

§ 4. Measurements of Gas Composition

in a Sealed CO_2 Laser

Interest in sealed lasers has recently increased because of the convenience of such light sources. The main shortcomings of such lasers are due to two factors. Firstly, the output power is somewhat lower than for a continuous-flow system and, secondly, the service life is limited unless the active medium is exchanged; this is undoubtedly due to chemical changes. In addition to the difficulties encountered in the analysis of chemical kinetics under continuous-flow conditions, in the case of sealed lasers we have to allow also for the interaction between the active gaseous phase and the enclosure. This is demonstrated forcefully by the observation that the service life of a sealed laser depends greatly on the nature of the tube and electrode materials.

Much work has been done on the interaction of atoms and molecules in gaseous phases with solids. However, an analysis of the published work gives only sufficient information for a preliminary choice of the material because of the special nature of the conditions in a discharge and because of the presence of different components which can be neutral or ionized [47]. This and the presence of excited states may result in such interactions in an electrical discharge which are unlikely under normal equilibrium conditions. The important factors are the cleanness of the surfaces and their preliminary treatment. A special feature of electrical discharges is also the fact that the electrode metal (particularly the cathode material) is sputtered during operation and this increases considerably the active surface. The intensity of sputtering of metals in discharges depends on many factors [48]. These factors include the discharge current, cathode voltage drop, electrode material, and composition of the gas. These and many other considerations have suggested that it is worthwhile studying separately the chemical reactions which occur in sealed CO_2 lasers under different conditions.

Experimental Method

We investigated sealed CO_2 lasers by the absorption of infrared radiation. In this case, the discharge gap was a gas-analyzer cell fitted with electrodes located in side tubes. The absorption measurements were carried out immediately after switching off a discharge. The discharge was then switched on for some time until the next measurement. The reliability of the results obtained in this way was checked as follows. In one experiment, the measurements were carried out after t minutes of operation of a discharge and, in another, they were carried out on the same initial mixture when the discharge was switched off twice after times t_1 and t_2 so that $t_1 + t_2 = t$. The results obtained in these two cases were identical, so we assumed that the composition of the mixture was "frozen" by switching off the discharge.

We also used a different method. A discharge tube of a laser was connected in parallel with a measuring cell of a gas analyzer so that the whole system was isolated. A glass piston pump with a valve system ensuring fast circulation of the mixture (~ 1 m/sec at the working pressures) was located in one of the connecting tubes. This made it possible to study the changes in the CO_2 and CO concentrations as a function of time in parallel with the discharge. In this way, we established that when a discharge was switched off the dependence was "frozen" but it continued during the subsequent operation. This confirmed additionally that the process of switching on a discharge did not give rise to significant changes in the time dependence of the concentrations of the products of the dissociation process.

A system of connecting tubes allowed us to circulate the gas mixture in any direction, i.e., the mixture leaving the discharge was passed directly to the measuring cell through a connecting tube (~ 30 cm) or through a system of pump tubes (~ 2 m). This corresponded to different times from the moment of emergence from the discharge zone to the moment of arrival in the measuring cell (these times were ~ 0.3 and ~ 2 sec, respectively). The results obtained in both cases were identical, which provided additional confirmation that the concentrations of CO_2 and CO did not change when the gas emerged from the discharge zone.

However, the second method was less convenient in the main measurements because it involved the participation of large inactive objects and surfaces. In addition, it gave rise to further difficulties in the interpretation of the time dependences of the concentrations of the components (particularly during the initial stage) because of the thermal expulsion of the gas from the discharge zone into the measuring cell. Therefore, the results reported below were obtained by the first method.

Experimental Results

Tube with Molybdenum Electrodes. We first studied in detail the changes in the gas composition in a tube made of molybdenum glass and fitted with hollow cylindrical molybdenum electrodes. The internal diameter of this tube was 30 mm and the length of the discharge gap was 460 mm.

Figure 15 shows the dependences of the CO_2 and CO concentrations on the duration of a discharge up to 3 h in CO_2-He (1:9) and CO_2-N_2-He (1:3:6) mixtures. On the whole, the dependences are the same except that the CO_2 in the ternary mixture decomposed more than in the binary mixture, exactly as in the continuous-flow case. The time dependences of the CO_2 and CO concentrations are nonmonotonic and show considerable changes (loss of CO_2 and increase in CO) during initial moments (this stage is not shown in Fig. 15 and it amounts to several seconds). The results are in good agreement with the observations made in the continuous-flow case that the chemical equilibrium of the dissociation reaction is established rapidly in a discharge. Subsequently (up to 20-40 min later), the concentration of CO_2 rises slowly and, after passing through a maximum, begins to fall slowly. It is very important to note, as indicated by Fig. 15, that there is no carbon balance. By way of control, we carried out analyses in discharges with pure N_2 and O_2 gases. When a discharge occurs in pure oxygen, we observe the appearance of a considerable amount of CO_2 (up to 15% compared with the concentration of CO_2 in the working mixtures). After a similar experiment with N_2 in the discharge tube, we observe no CO_2 but a small amount of CO. We mentioned earlier the cause of this effect (see § 3), which was the oxidation of CO and C present on the electrodes and the discharge-tube walls.

The fall in the CO_2 concentration (after 20-40 min) could be due to the fact that oxygen was lost from the discharge zone because of the adsorption (or chemisorption) on the tube walls and on the electrodes, as established in [14] for aluminum electrodes. A comparison of the results reported in [14] with those given in the present paper indicated that in our case the loss of CO_2 (including the rapid loss during the initial moments) was much slower. In fact, our experiments indicated that, other conditions being equal, stimulated emission from a laser with Mo electrodes lasted much longer than from one with Al electrodes.

Figure 16 shows the dependences of the CO_2 and CO concentrations on the duration of a discharge during initial moments (up to 5 min). Once again, there was no carbon balance. The

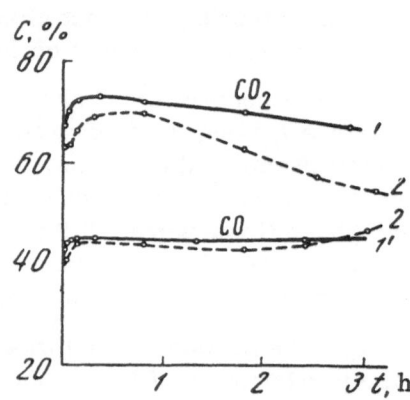

Fig. 15. Dependences of the CO_2 and CO concentrations on the duration of a discharge in a sealed system: 1, 1') CO_2-He mixture; 2, 2') CO_2-N_2-He mixture. The tube had molybdenum electrodes; p = 4.1 Torr; i_d = 30 mA.

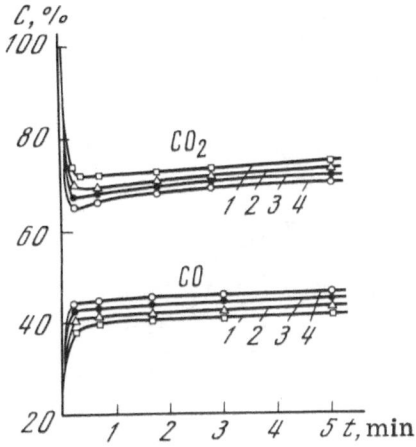

Fig. 16. Dependences of the CO_2 and CO concentrations on the duration of the initial stages of a discharge in a CO_2-N_2-He ($1:3:6$) mixture plotted for different discharge currents i_d (mA): 1) 10; 2) 20; 3) 30; 4) 40. The tube had molybdenum electrodes; p = 4.1 Torr.

dependences of the CO_2 and CO concentrations on the time spent by the mixture in the discharge varied little when the current was increased from 10 to 40 mA, in good agreement with the results obtained for continuous-flow systems. It is clear from Fig. 5 that a reduction in the rate of flow shifted the main part of the dependence of the CO_2 concentration on the discharge current to the low-current range (below 10 mA).

The information obtained on sealed CO_2 lasers can explain some of the experimental observations reported earlier for CO_2 lasers. For example, the output power of a sealed laser passed through a maximum after 30-40 min operation, which could be explained by the curves in Fig. 15. Cheo and Cooper [49] carried out an investigation with a satisfactory time resolution and found that the output power and gain varied rapidly during the first moments after beginning of a discharge and this could be explained by the curves in Figs. 15 and 16. Moreover, in contrast to systems with fast gas flow, the output power from sealed pure CO_2 lasers was comparable with that obtained from a sealed CO_2-N_2 laser [50]. This was due to the fact that the initial rapid dissociation of CO_2 into CO and O and the subsequent relative stability of the concentrations of these products gave rise to population inversion as a result of transfer of the vibrational energy from CO to CO_2.

A sealed CO_2 laser ceased to work because of the irreversible loss of the CO_2 dissociation products to the electrodes and discharge-tube walls. Hence, we could draw a conclusion of practical importance: The service life of a sealed laser could be increased either by reducing the rate of irreversible adsorption processes (by the selection of suitable materials) or by reducing the degree of dissociation of CO_2 (by the introduction of some catalysts favoring oxidation of CO back to CO_2). We investigated how these two possibilities worked in practice.

Tubes with Electrodes of Other Metals. We studied changes in the CO_2 concentration in discharge tubes which had copper, nickel, tantalum, and platinum electrodes. The aim was to determine not only the time dependences of the concentration in one experiment but also the reproducibility of the results. A tube was washed with alcohol and then pumped for a long time to a pressure of $\sim 10^{-3}$ Torr. In all cases, a discharge occurred in a CO_2-N_2-He ($1:3:6$) mixture with an initial pressure of 6.9 Torr.

Figure 17 shows the dependences of the CO_2 content (in percent) relative to the initial amount in a tube with copper electrodes. There are six curves in this figure. Curves 1-6 were obtained in a sequence of measurements under identical conditions. The results differed very strongly. When the tube was first filled with a mixture and a discharge was switched on, a rapid (lasting several seconds) fall in the number of CO_2 molecules was followed by a rise so that, after 20 min, the amount of CO_2 exceeded the initial value. The tube was then pumped

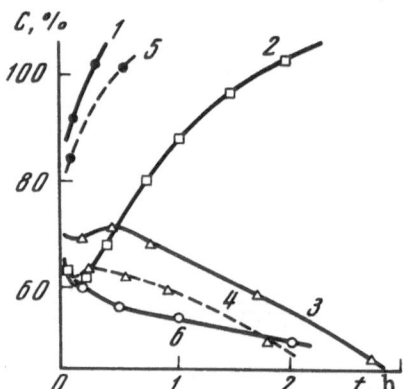

Fig. 17. Dependences of the CO_2 concentration on the duration of a discharge in a CO_2-N_2-He (1:3:6) mixture at p = 6.9 Torr for i_d = 30 mA.

out and the experiment repeated under identical conditions (curve 2). Once again, a rapid dissociation was followed by a rise in the CO_2 concentration although this was slower than in the first case. Curves 3, 4, and 6 demonstrated a dependence analogous to that described above for the Mo electrodes: After several minutes, CO_2 was desorbed and then the concentration fell strongly. However, in the case of molybdenum electrodes, the results were reproducible from one experiment to another, whereas in the case of copper electrodes there was no reproducibility and curve 5 behaved quite differently (it was similar to curves 1 and 2).

This behavior could be explained as follows. During the preparation of a discharge tube (welding of electrodes, sealing them into the glass, and joining various glass parts of the tube), oxygen and carbon compounds were adsorbed on the electrodes and the tube walls; these adsorbed substances could not be removed by simple washing followed by evacuation. A discharge (involving an increase in the temperature as well as electron and ion bombardment) caused desorption of these compounds, which increased the CO_2 concentration, as described by curves 1 and 2. The rate of this process was considerably less than the rate of dissociation in the plasma. After several hours of operation of the discharge and pumping of the gas, the amounts of these compounds on the surfaces decreased and subsequently they could not alter significantly the equilibrium CO_2 concentration in the plasma.

Even slower adsorption (or chemisorption) processes were responsible for the subsequent fall of the CO_2 concentration. For example, copper oxidized rapidly at ~130°C [51] and the loss of oxygen shifted the equilibrium of the reaction (1) to the right. These processes were responsible for curves 3, 4, and 6; moreover, the reduction in CO_2 from one experiment to another became increasingly greater when the discharge was operated for several hours. This was due to the strong sputtering of copper, noticeable even after 2–4 h operation. The exception was curve 5 but it probably represented the sudden opening of a macroscopic inclusion in the metal as a result of sputtering.

Figure 18a shows similar curves obtained for a tube with nickel electrodes. After several minutes of the first run, the carbon compounds evolved from the electrodes restored the CO_2 concentration to its initial value and then the amount of carbon dioxide began to fall quite strongly. This behavior could be explained as follows. In this case, it was important which components were first adsorbed by nickel. Kawasaki et al. [52] established that when CO was adsorbed first and then the surface of the nickel was brought into contact with oxygen, the latter reacted with CO and this caused desorption of CO_2. When oxygen was adsorbed first, it did not react with CO, i.e., CO_2 was formed neither in the gaseous phase nor in the bound form in the metal.

Thus, the rise of CO_2 during the first minutes was due to the presence of CO and CO_2 adsorbed during the construction of the discharge tube and the fall in concentration of the

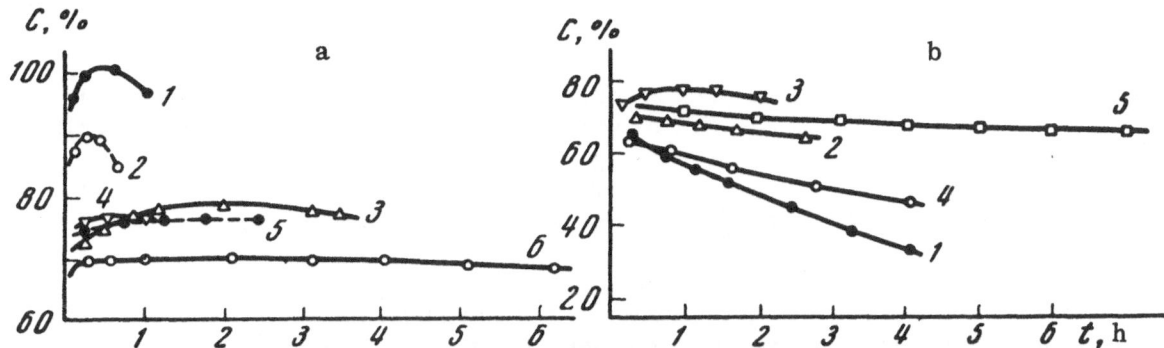

Fig. 18. Dependences of the CO_2 concentration on the duration of a discharge in a CO_2- N_2-He (1:3:6) mixture at 6.9 Torr for $i_d = 30$ mA: a) tube with nickel electrodes; b) tube with tantalum electrodes; curves 1-6 represent consecutive runs.

carbon dioxide was due to the loss of oxygen from the discharge zone (curves 1 and 2). The metal surfaces became saturated with oxygen when the discharge operated for several hours and these surfaces no longer greatly affected the equilibrium of the reaction (1), as demonstrated by curves 3-6. Nevertheless, the steady-state concentrations of CO_2 decreased from one run to another (other conditions being equal). This was due to the sputtering of nickel in the discharge.

An investigation of the possibility of using nickel electrodes in sealed CO_2 lasers was also carried out by Carbone [25, 26] and by Deutsch and Horrigan [53]. They found that the stimulated emission lasted much longer when the electrodes were kept at an elevated temperature ($\sim 300°C$). The explanation provided in [25, 26] was that heated nickel acted as a catalyst of the $CO + O \to CO_2$ reaction. Deutsch and Horrigan [53] assumed that the heating of electrodes prevented the formation of an NiO film on the surface and even when the tests were initially carried out using cold electrodes the fall in the output power could be reversed to a rise by heating these electrodes.

In contrast to copper and nickel, tantalum electrodes experienced no significant sputtering even after several tens of hours of discharge operation. The results obtained using a tube with tantalum electrodes are plotted in Fig. 18b. Before filling with a working mixture of gases, we conditioned the tube by a xenon discharge lasting 10 h. It is clear from Fig. 18b that, in the first discharge in the working mixture (curve 1), the amount of CO_2 rapidly fell during the first 3-4 h. The tube was then pumped out and again tested under the same conditions (curve 2). In this case, the loss was slower than in the first case and even slower during the next run (curve 3). The conditioning in xenon for 10 h was then repeated and new tests were carried out. Once again, there was a rapid reduction in the amount of CO_2 (curve 4), which slowed down during subsequent runs (curve 5). The results obtained could be explained very simply. Conditioning by a xenon discharge cleaned the surface of tantalum and this increased its adsorptivity. During subsequent runs, the tantalum surface became saturated and the adsorption slowed down because of the weak sputtering of this metal.

It seemed very desirable to investigate the behavior of a tube with platinum electrodes. Firstly, this was a noble metal with an extremely weak reactivity and, secondly, it exhibited a catalytic activity in oxidation processes [41]. The most successful sealed CO_2 lasers were obtained by the present author and by others [54] when platinum electrodes were used.

The results were as expected (Fig. 19). After the first two runs (curves 1 and 2), during which the CO_2 concentration increased because of insufficient preliminary cleaning of the surface, the results were highly reproducible (to within the experimental error, estimated to be

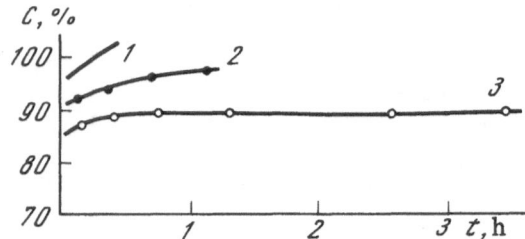

Fig. 19. Dependences of the CO_2 concentration on the duration of a discharge in a CO_2-N_2-He $(1:3:6)$ mixture at 6.9 Torr for $i_d = 30$ mA. The tube had platinum electrodes; curves 1-3 represent consecutive runs.

2-3%). Curve 3 represents the results of one of the many runs, including those carried out after prolonged conditioning in a xenon discharge. The most important observation was that the dissociation of CO_2 was weak (~10%) and much less than for any other electrodes.

Many workers (see, for example, [55]) have recently started using xenon as an additive to the working gas; this has made it possible to increase the efficiency by 15-20%. Moreover, evidence is available [56] that the addition of xenon could prolong the service life of a sealed laser. In view of this, we studied the influence of xenon on the dissociation of CO_2 in tubes with nickel and tantalum electrodes. However, we found no significant influence of xenon (at least during the first few hours of operation).

Influence of Hydrogen and Water Vapor Admixtures. An increase in the output power of a sealed CO_2 laser as a result of the addition of a small amount of water vapor to a CO_2-N_2 working mixture was first reported by Witteman [57]. For example, the introduction of about 0.1 Torr H_2O vapor doubled the output power. Witteman attributed the increase in output power to an effective deactivation of the lower laser level as a result of the CO_2-H_2O collisions. However, although the efficiency of this process was known to be very high, this interpretation could not be regarded as final. This was due to the fact that the addition of hydrogen and water to a discharge could also alter considerably the molecular composition of the plasma. For example, the earliest studies of Fischer et al. [58] indicated that the degree of dissociation of moist carbon dioxide was less than that of the dry substance. .The same conclusion was reached by Reeves et al. [59] in later experiments involving the photolysis of CO_2.

In view of this situation, we measured the degree of dissociation of CO_2 in a sealed system in the presence of small amounts of hydrogen in a tube fitted with molybdenum electrodes. We used hydrogen because, on the one hand, Rosenberger [60] established that hydrogen and water had a similar influence on the laser properties and, on the other, the amount of the additive could be more easily determined in the case of hydrogen.

Hydrogen was introduced into the system by electric-current heating of a tantalum absorber.*

Figure 20 shows the dependences of the CO_2 concentration on the duration of a discharge. It is clear from this figure that the addition of 0.14 Torr H_2 reduced considerably the degree of dissociation of CO_2. Figure 21 shows the dependence of the CO_2 concentration after a 2-min discharge (steady-state dissociation) on the amount of hydrogen added. We found that a considerable reduction in the degree of dissociation of CO_2 occurred only for small amounts of hydrogen (up to 0.2-0.3 Torr). A further increase in the concentration of H_2 had little effect on the dissociation of carbon dioxide.

---

*An analysis of complex gas mixtures, carried out by the chromatographic method, demonstrated that the amount of moisture in these mixtures did not exceed 0.01% of the total pressure in the mixture.

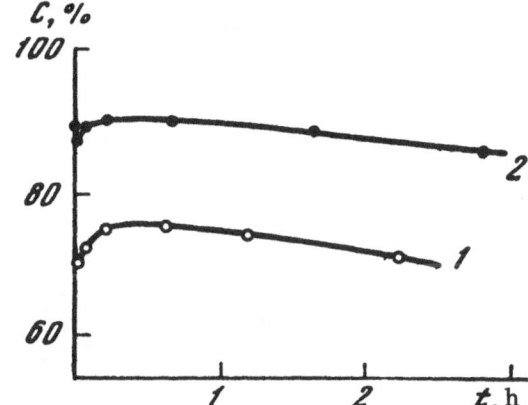

Fig. 20. Influence of hydrogen on the dissociation of CO_2 in a discharge in a tube with molybdenum electrodes: 1) without H_2; 2) with H_2.

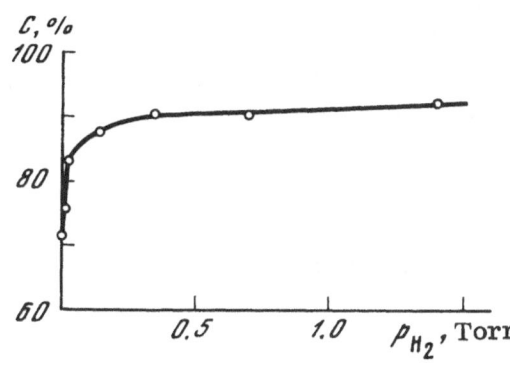

Fig. 21. Dependence of the dissociation of CO_2 after a 2-min discharge on the amount of hydrogen added to a tube with molybdenum electrodes.

These results demonstrated that small amounts of hydrogen (or water vapor) could increase the output power and service life of sealed CO_2 lasers not only because of relaxation processes associated with the deactivation of the lower active level but also because of a reduction in the degree of dissociation of CO_2 [28].

The same conclusions were reached later by Witteman [54], Lotkova et al. [61], Almer et al. [37], and Karube et al. [62], who also studied the dissociation of CO_2 in laser discharges in the presence of hydrogen. Figure 22 gives the results obtained by Witteman [54], who used the mass spectrometric method. It is clear from Fig. 22 that the degree of dissociation of CO_2 was low when hydrogen was present in the discharge. The rapid fall of CO_2 and rise of CO were paralleled by the loss of hydrogen and a fall of the output power.

The relationship between the relaxation and chemical effects of H_2 and H_2O on the output power is not clear. It is desirable to know this relationship because it would help in a better understanding of the population inversion mechanism. According to the available data on the probability of deactivation of vibrations, the lower laser level may be depopulated quite effectively also in the absence of H_2 and H_2O, particularly in the presence of He in the laser mixture (this point is discussed in detail later). If a strong influence of H_2 and H_2O on the population of the lower laser level were to be established firmly, it would be necessary to review the published data on the relaxation processes.

Deutsch and Horrigan [53] compared the results of other workers with their own on the use of H_2 and H_2O and concluded that, when the partial pressure of CO_2 in a sealed laser was low, the effect of H_2 and H_2O was much greater than at high CO_2 pressures. In continuous-flow systems, the role of these additives decreased with increasing rate of flow. This undoubtedly indicated a considerable effect of H_2 and H_2O as additives which reduced the degree of dissociation of CO_2 because, as reported earlier, in the absence of H_2 and H_2O, the degree of dissociation of CO_2 decreased with rising pressure and rate of flow.

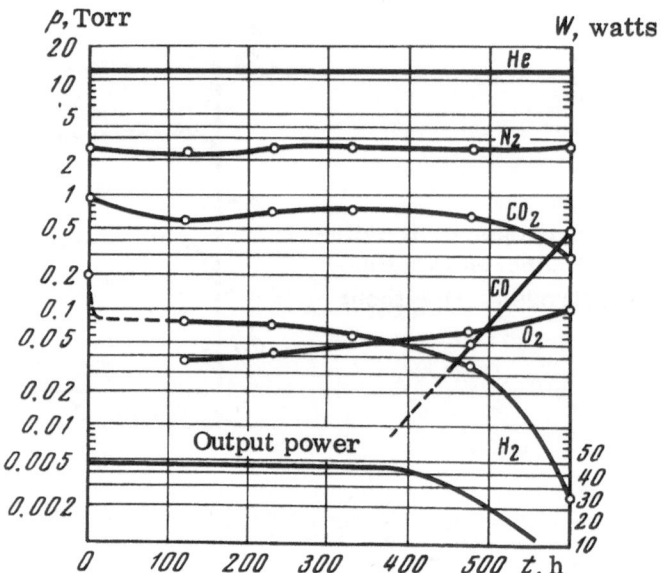

Fig. 22. Changes in the composition of the components of a mixture during a discharge in a sealed laser with a long service life [54]. The tube diameter was 20 mm and the electrodes were made of platinum.

The concentration of the hydroxyl radical OH in the discharge was measured in order to obtain more detailed information on the influence of H_2 and H_2O on the composition of the gas mixture. The measurements were carried out using the Kondrat'ev method [63]. The concentration was determined by recording the changes in the intensities of the OH lines (produced using an auxiliary source) transmitted by the investigated discharge gap. We found that for typical (in lasers) amounts of water vapor (0.1-0.2 Torr) and discharge currents (10-40 mA) the density of the OH radicals was $(2-5) \cdot 10^{14}$ cm^{-3}, i.e., the degree of dissociation of water vapor in the discharge did not exceed 10%. It was interesting to note that, within the limits of the experimental error ($\pm 30\%$), the density of the OH radicals was unaffected by the presence of H_2O vapor or H_2. In spite of the fact that the density of the OH radicals was relatively low, their high reactivity could alter considerably the rate of oxidation of CO to CO_2: CO + OH $\rightarrow$ CO_2 + H, resulting in a reduction in the steady-state degree of dissociation of carbon dioxide in a discharge.

We shall conclude by pointing out that a comprehensive study of the influence of H_2 and H_2O admixtures on the laser action should include an investigation of the relaxation processes involving the H atoms and OH radicals. Moreover, even small amounts of H_2 and H_2O can alter the electron velocity distribution in a discharge.

CHAPTER II

RELAXATION PROCESSES AND POPULATION INVERSION IN A MULTICOMPONENT PLASMA IN A CO_2 DISCHARGE LASER

§ 1. Introduction

The investigation of the dissociation of CO_2 reported in the preceding chapter shows clearly that, from the chemical point of view, the CO_2 laser plasma is a multicomponent mixture. This must be borne in mind in studies of the CO_2 laser action, i.e., in discussing the processes which ensure pumping of the upper laser level and the relaxation processes which depopulate the lower and upper laser levels. The experimental results and preliminary theoretical considerations suggest that the CO_2, N_2, and CO molecules play the dominant role in the CO_2 laser action. Therefore, before considering in detail the population inversion mechanism in a CO_2 laser, we must recall some fundamental information on these molecules, consider some aspects of the theory of relaxation in a mixture of molecular gases, and present briefly the available information on the CO_2 laser action.

Molecules of CO_2, N_2, and CO in Their Ground Electronic States

Infrared radiation is emitted from CO_2 lasers as a result of stimulated transitions between the vibration-rotational levels of the CO_2 molecule belonging to the ground electronic state. The CO_2 molecule is linear and symmetric (point symmetry group $D_{\infty h}$). The lowest vibrational levels of this molecule are shown schematically in Fig. 23.

The cw emission is due to the 00^01–10^00 and 00^01–02^00 transitions. Since the upper laser level 00^01 belongs to the antisymmetric vibration with a dipole moment directed along the molecule axis, the general selection rule $\Delta J = 0 \pm 1$ is supplemented by the condition $\Delta J = 0$ and the Q branch is absent from the stimulated emission spectrum so that only the parallel R- and P-branch bands are observed. Transitions from the 01^10 level, which belongs to the system of energy levels of the deformation vibrations, give rise to transverse bands. The selection rule for l in the parallel bands is $\Delta l = 0$ so that there are no lines due to the 00^01–02^20 transitions.

The experimental values of the probabilities of some of the radiative transitions between the vibrational levels of the CO_2 molecule can be found in Penner's book [64] and in later investigations carried out by the laser method [65, 66]. Moreover, some transition probabilities are calculated in [67]. The strongest transitions with $\Delta v_1 = \Delta v_2^l = 0$ and $\Delta v_3 = 1$ have probabilities of $\sim 10^2$ sec^{-1}.

Figure 23 gives not only the CO_2 levels but also the lower vibrational levels of the CO and N_2 molecules. The probabilities of transitions in the CO molecule in the ground electronic

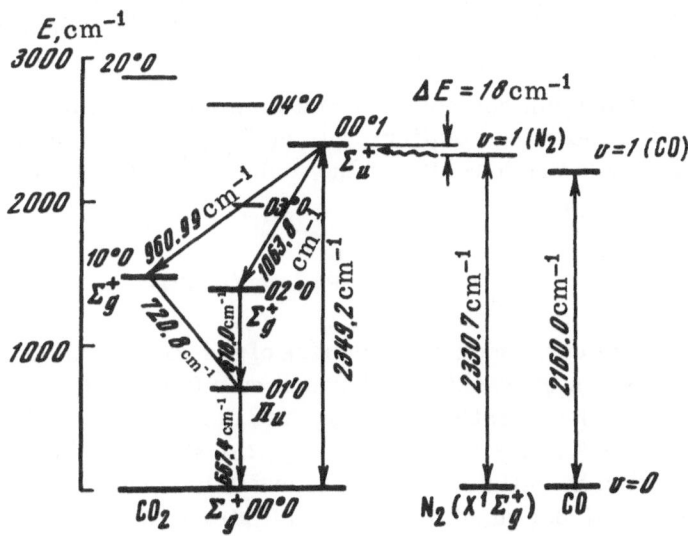

Fig. 23. Lower vibrational levels of the CO_2, N_2, and CO molecules.

state $^1\Sigma^+$ are ~10 sec^{-1}, whereas the infrared radiation of the N_2 molecule (ground state $^1\Sigma_g^+$) is symmetry-forbidden in the dipole approximation.

Relaxation Processes in Multicomponent Gas Mixtures

The departure of a system from equilibrium gives rise to relaxation processes which tend to restore the full statistical equilibrium.* If a relaxing gas has a nonequilibrium distribution for any one degree of freedom, the relaxation process in this very simple case is described by

$$\frac{dE(t)}{dt} = \frac{E(t) - \bar{E}}{\tau}, \tag{2.1}$$

where E(t) is the energy of this degree of freedom and $\bar{E}$ is the equilibrium value of the energy.

Vibrational Relaxation. In the case of a laser utilizing transitions between the vibrational levels of the CO_2 molecule, the populations of the active levels differ strongly from the equilibrium values. Therefore, the relaxation processes tending to reestablish the equilibrium in respect to the vibrational degrees of freedom should play an important role and have a considerable influence on the CO_2 laser action. We shall now consider such relaxation processes.

The simplest case of vibrational relaxation is that which occurs in a system of diatomic molecules which are present as a small admixture to an inert monatomic gas so that the interactions between these molecules can be ignored. In this case, the relaxation is described by Eq. (2.1) with a single time constant τ representing the process of reestablishment of the vibrational equilibrium. This is also true of the vibrational relaxation when the concentration of molecules of one kind in an inert gas is not low and we have to allow for the exchange of vibrational quanta between these molecules. The physical meaning of this result is that the rate of vibrational relaxation is governed only by the processes involving the creation and annihilation of vibrational quanta. The rate of these processes depends only on the efficiency of energy exchange between the translational and vibrational degrees of freedom of the colliding molecules. The exchange of vibrational quanta has no effect on the rate of relaxation because it simply redistributes the existing quanta. The situation is different if we consider the vibrational relaxation in a mixture of polyatomic gases because the processes of redistribution of vibrational energy between different components may alter the relaxation time of the mixture as a whole.

We shall consider a system composed of diatomic molecules of different kinds (A and B) and we shall introduce the following notation: $h\nu_A$ and $h\nu_B$ are the vibrational quanta of the two molecules; $x_n(t)$ and $y_n(t)$ are the densities of the molecules A and B at the n-th vibrational level, where

$$\sum_{n=0}^{\infty} x_n(t) = N_A, \qquad \sum_{n=0}^{\infty} y_n(t) = N_B; \tag{2.2}$$

Z_{AA}, Z_{AB}, and Z_{BB} are the corresponding numbers of gas-kinetic collisions; $\mathscr{P}_{nm}(AA)$ and $\mathscr{P}_{nm}(AB)$ are the probabilities of transition of a molecule from a state n to a state m accompanied by the conversion of the vibrational energy into the energy of translation motion as a result of collisions with the A and B molecules; $\mathscr{P}_{nm}(BB)$ and $\mathscr{P}_{nm}(BA)$ are the corresponding probabilities for a molecule B colliding with the B and A molecules; $Q_{n,\,n+l}^{m+k,\,m}$ is the probability of the transfer of a vibrational quantum by an oscillator in a state m + k to an oscillator

---

*A detailed theory of relaxation can be found, for example, in monographs [68-70].

in a state n. The probabilities $\mathscr{P}$ and Q obey (in the harmonic oscillator approximation) the relationships

$$\mathscr{P}_{n+1,\,n}(AA) = (n+1)\,\mathscr{P}_{10}(AA),$$

$$\mathscr{P}_{n+1,\,n}(AB) = (n+1)\,\mathscr{P}_{10}(AB), \qquad (2.3)$$

$$Q^{m+1,\,m}_{n,\,n+1}(AB) = (m+1)(n+1)\,Q^{10}_{01}(AB),$$

where N is the total number of molecules.

Adopting this notation, we obtain the following system of equations, which describes the balance of the densities of molecules in different vibrational levels:

$$\frac{dx_n}{dt} = Z_{AA}[\mathscr{P}_{n+1,\,n}(AA)\,x_n - \mathscr{P}_{n,\,n+1}(AA)\,x_{n+1}]\frac{N_A}{N} +$$

$$+ Z_{AB}[\mathscr{P}_{n+1,\,n}(AB)\,x_{n+1} - \mathscr{P}_{n,\,n+1}(AB)\,x_n]\frac{N_B}{N} +$$

$$+ Z_{AB}\sum_m [Q^{m,\,m+1}_{n+1,\,n}(AB)\,x_{n+1}y_m - Q^{m+1,\,m}_{n,\,n+1}(AB)\,x_n y_{m+1}]\frac{1}{N}, \qquad (2.4)$$

which should be supplemented by similar equations for dy_n/dt.

These balance or rate equations for the vibrational energy can be obtained by summing the expressions (2.4) over all the vibrational levels, applying the selection rules (2.3) and the principle of detailed equilibrium $\mathscr{P}_{10} = \mathscr{P}_{01}e^{-\vartheta}$, $Q^{10}_{01} = Q^{01}_{10}\exp(\vartheta_A - \vartheta_B)$, where $\vartheta = h\nu/kT_g$, ν is the frequency of a vibrational quantum, and T_g is the gas temperature. If we use E' to denote the dimensionless density of the vibrational energy (i.e., the total number of vibrational quanta per unit volume),

$$E'_A = \frac{E_A}{h\nu_A} = \sum_{n=0}^{\infty} n x_n(t),$$

we find that the relaxation equations for the dimensionless density of the vibrational energy become

$$\frac{dE'_A}{dt} = -\frac{1}{\tau_A}(E'_A - \bar{E}'_A) + \frac{1}{\tau_{AB}}[E'_B\exp(\vartheta_B - \vartheta_A)(E'_A + N_A)] - E'_B(E'_B + N_B)], \qquad (2.5)$$

which should be supplemented by a similar expression for dE'_B/dt. Here,

$$\tau_A = [Z_{AA}\mathscr{P}_{10}(AA)N_A + Z_{AB}\mathscr{P}_{10}(AB)N_B(1 - e^{-\vartheta_A})]^{-1},$$

$$\tau_{AB} = [Z_{AB}Q^{01}_{10}(AB)]^{-1},$$

where $\bar{E}'_A = N_A/[\exp(\vartheta_A) - 1]$ is the equilibrium value of the vibrational energy.

Equation (2.5) differs from Eq. (2.1) by the second term, which allows for the exchange of vibrational quanta between the A and B molecules.

Expression (2.5) simplifies if the A and B molecules have similar energies of the vibrational quanta, i.e., if $\vartheta_A \approx \vartheta_B$. Bearing in mind that $N_A/N_B = (\bar{E}'_A/\bar{E}'_B)[(e^{\vartheta_A} - 1)/(e^{\vartheta_B} - 1)] \approx \bar{E}'_A/\bar{E}'_B$, we obtain

$$\frac{dE'_A}{dt} = k_T\left(\frac{\bar{E}'_A}{\bar{E}'_B}E'_B - E'_A\right) - k_L(E'_A - \bar{E}'_A), \qquad (2.6)$$

where $k_L = \tau_A^{-1}$ and $k_T = \tau_{AB}^{-1}$ are the rates of transfer of the vibrational energy to the translational degrees of freedom and of the exchange, respectively. We can solve the system (2.6) if we know the values of k_L and k_T related to the probabilities $\mathscr{P}_{10}$ and Q_{10}^{01}. These probabilities can be calculated by solving the Schrödinger equation for the collision process. Such calculations are reported for specific cases in [68, 71, 72]. The precision of these calculations is only to within one order of magnitude. This is primarily due to the simplifications made in the choice of the intermolecular interaction potential.

An important aspect of the vibrational relaxation theory is the temperature dependence of the relaxation probability. For example, the dependence ln $\mathscr{P}_{nm} \propto T^{1/3}$ obtained in [73] is supported by the experimental results (see, for example, [74]). However, in some cases, the temperature dependence can be described by the Arrhenius law $\mathscr{P}_{nm} = Ae^{-B/T}$ [75-77]. Calculations reported in [68, 71] do not yield the analytic temperature dependence of $\mathscr{P}_{nm}$. However, these calculations can be used to find values of $\mathscr{P}_{nm}$ at various temperatures. Such a calculation shows that at high temperatures the dependence $\mathscr{P}_{nm}(T)$ is stronger than at low temperatures and it is close to $\mathscr{P}_{nm} = AT^r e^{-B/T}$. It also follows from these calculations that the probability is $Q_{n,\,n+l}^{m+k,\,m} \propto T$.

In view of these points, the temperature dependence of the probability should be regarded as tentative and requiring refining calculations and additional experiments.

The dependence of $\mathscr{P}_{nm}$ on the quanta transformed into the translational energy is [73]

$$\mathscr{P}_{nm} \propto \nu_{nm} e^{-\beta \nu_{mn}^{2/3}}, \qquad (2.7)$$

where $\beta = 3[2\pi^4 a^2 (m/kT)]^{1/3}$; $a \sim 10^{-8}$ cm is the radius of the intermolecular interaction. An analysis of the above expression shows that, in all cases of practical interest, the value of $\mathscr{P}_{nm}$ is a falling function of the quantum energy.

Establishment of Maxwellian Distribution. Rotational Relaxation. If τ_t is used to denote the time for the establishment of the Maxwellian distribution between the translational degrees of freedom, a simple analysis based on the hard-sphere model shows that $\tau_t \sim 1/z \equiv \tau_0$, where τ_0 is the time between gas-kinetic collisions and z is the frequency of these collisions.

Similarly, we can show that if $kT_g > B_v$ (B_v is the rotational constant), the rotational relaxation time τ_r is of the order of the translational relaxation time although slightly larger than the latter. A comparison of the characteristic time for the establishment of the Maxwellian distribution of molecules between the translational degrees of freedom (τ_t), time for the establishment of the Boltzmann distribution between the vibrational levels (τ_r), and vibrational relaxation time (τ_v) shows that, in most cases of practical interest (including the discharges studied in the present paper), the following relationship is satisfied:

$$\tau_t \lesssim \tau_r \ll \tau_v. \qquad (2.8)$$

It follows that in solving the problem of the vibrational relaxation in a CO_2 laser, we can use Eq. (2.6) on the assumption that at each moment during vibrational relaxation there is an equilibrium in respect of the translational and rotational degrees of freedom.

It follows that, in principle, it should be possible to describe accurately the process of vibrational relaxation by Eq. (2.6) if we know the probabilities of exchange of vibrational quanta and of the conversion of these quanta into the energy of translational motion as a result of collisions between molecules. However, we can see from the foregoing discussion that these two probabilities cannot yet be calculated theoretically with sufficient accuracy. Therefore, in a quantitative solution of the relaxation problem, we have to use experimental values of the probabilities of elementary processes.

Population Inversion in a CO_2 Laser

Sobolev and Sokovikov [5, 9, 11] developed the basic physical ideas on the mechanism of population inversion in a CO_2 laser based on the experimental work of Schulz [10] and Swift [78], who determined the excitation cross sections of vibrations of the N_2 and CO molecules by electron impact and investigated the electron velocity distribution in a discharge in N_2. A strong "overlap" of the excitation cross sections and distribution functions along the energy scale led them to conclude that the electron excitation mechanism predominated and that it was capable of ensuring the necessary concentrations of excited molecules.

Thus, the upper CO_2 laser level is pumped mainly by electron impact whereas the main process responsible for the population inversion is the deactivation (depopulation) of the lower level as a result of molecular collisions. A method for the quantitative calculation of the populations of the vibrational levels in a CO_2 laser is put forward in [12] on the basis of the ideas developed in [5, 9, 11]. This method is based on the kinetics of population and depopulation of the levels. In principle, the calculations can be made by solving a set of rate equations for a multilevel system as a function of many parameters such as the particle concentration, discharge current, gas pressure, etc. In view of the rich system of energy levels of the molecules participating in the population inversion, it is not possible to solve this problem completely. The main simplifying assumption suggested is the introduction of vibrational temperatures for each normal mode. This is possible because of a strong collisional coupling within a given mode. An assumption of this kind allows the replacement of a large number of rate equations for the individual levels with a system of a much smaller number of equations for each group of levels.

Since the gas temperature is an important factor that exerts a considerable influence on the vibrational level relaxation (this point has been discussed earlier), the rate equation system should be supplemented by the heat conduction equation. The proportion of the discharge energy used to heat the gas is then defined as the energy transferred by relaxation from the vibrational to the translational degrees of freedom.

The results obtained in [12] describe correctly the qualitative nature of the experimental dependences of the population inversion density on various discharge parameters. However, quantitative results are only in order-of-magnitude agreement with the experimental results. This discrepancy is due to the fact that although the approach adopted in [12] is correct, many important factors have been ignored.

A serious shortcoming of the calculations in [12] is the lack of allowance for changes in the gas composition under the action of a discharge. The possibility of electron-impact excitation of the vibrations of CO_2 molecules and their dissociation products is not allowed for and it is assumed that only N_2 is excited. Considerable sources of error may arise also from the unjustified selection of the nature of the electron velocity distribution function (Maxwellian) and from the rough estimate of the electron density and gas temperature.

The equations employed in [12] to determine the vibrational temperatures are essentially the rate equations for the lower vibrational levels so that the role of the higher levels is ignored. Moreover, some of the probabilities of the relaxation processes and their temperature dependences are calculated in [12] with limited precision. Our task will be to consider the kinetics of physical processes in a CO_2 laser and to calculate the population inversion density primarily in order to determine the role of dissociation of CO_2 in a discharge. We shall try to allow for the other factors mentioned above and use the most reliable experimental data on the probabilities of elementary processes and discharge parameters of a CO_2 laser.

§ 2. Derivation of Rate Equations

We shall derive the rate equations for the energy in various vibrational modes of molecules by considering the following physical processes. We shall assume that the excitation of the vibrations in the N_2 and CO molecules and of the antisymmetric vibrations of CO_2 is due to direct electron impact. The similarity of the energies of the lower levels ensures a resonant transfer of energy in collisions between the various molecules as well as exchange of vibrational quanta within each mode. The vibrational levels are deactivated by collisions with molecules, atoms, and electrons (mainly via the deformation vibrations of the CO_2 molecules) and with the discharge-tube walls. We shall ignore the direct transfer of the vibrational energy from the nitrogen and CO molecules and from the antisymmetric vibrations of CO_2 to the translational motion because the quanta in question are large. Simple estimates show that radiative decay of the levels under consideration can also be ignored [67].

We have already mentioned that the introduction of the concept of a vibrational temperature of each mode is justified only if the exchange of quanta within each mode is much faster than the arrival or loss of quanta (exchange, relaxation). However, it follows from theoretical considerations and recent experiments [79] that there is an intense energy exchange (mainly because of the Fermi resonance) between the symmetric and deformation vibrations. The rate of this exchange is comparable with the exchange within the vibrational modes. This not only complicates the picture but also allows us to introduce a unified vibrational temperature for both types of vibration. We shall assume that this temperature is approximately equal to the gas temperature. This assumption is justified by the following observations.

1. In working mixtures usually employed in CO_2 lasers, the relaxation of the energy of the antisymmetric vibrations of CO_2 (and of the vibrations of CO and N_2) to the energies of the symmetric and deformation vibrations is slower than the relaxation of the energy of the latter vibrations to the translational motion [80, 81].

2. Estimates given below indicate that the electron excitation of these vibrations also fails to make a significant contribution to the population of the levels under investigation.

3. During stimulated emission, the population of the lower laser level does not deviate significantly from the equilibrium value because of stimulated transitions even when the stimulated radiation density is in excess of 100 W/cm^2 [82].

It follows from the foregoing observations that the system of rate equations can be written in the form

$$
\frac{1}{V}\int \frac{dE_i}{dt}\,dV = h\nu_i \left\{ \alpha_i \left[N_{0i}k^{ei}\int n_e(r)\,dV - \right. \right.
$$
$$
\left. - k^{ie}\int N_i(r)\,n_e(r)\,dV \right] - \beta_i D_i \int \nabla^2 N_i(r)\,dV \left. \right\} \frac{1}{V} +
$$
$$
+ \sum_j k_T^{ji}\left[\frac{E_i}{E_j}E_j - E_i \right] - k_L^i [E_i - \bar{E}_i],
$$
$$
k_L^i = \sum_m \Psi_m k_L^{im}.
$$

(2.9)

Here, the indices i and j may assume the values 3, 4, and 5, denoting (in accordance with Fig. 23) the antisymmetric vibrations of CO_2 and the vibrations of N_2 and CO, respectively; ν is the vibration frequency; V is the volume of the discharge tube; N_{0i} is the number of oscillators of type i in the ground state (per unit volume); $k^{ei} = \langle v_e \sigma_i \rangle$ is the cross section for the excitation of i-type vibrations, averaged over the electron velocities; k^{ie} is the corresponding cross section for the deexcitation of the same vibrations; E_i is the energy of i-mode vibrations; $\bar{E}_i$ is the equilibrium value of the energy:

$$E_i = h\nu_i \frac{x_i}{(1-x_i)^2} N_{0i}, \quad x_i = \exp\left\{-\frac{h\nu_i}{kT_i}\right\},$$
$$\bar{E}_i = h\nu_i \frac{x_i}{(1-\bar{x}_i)^2} N_{0i}, \quad \bar{x}_i = \exp\left\{-\frac{h\nu_i}{kT_g}\right\};$$

(2.10)

T_i and T_g are the vibrational and gas (translational) temperatures, respectively; n_e is the electron density; N_i is the concentration of excited molecules; D_i is the diffusion coefficient; α_i represents the effective number of quanta per one excitation of a vibration by electron impact; k_T^{ji} is the rate of exchange of vibrational quanta in collisions between molecules; k_L^{im} is the rate of relaxation of the energy of i-mode vibrations as a result of collisions with molecules of the component m; Ψ_m is the fraction of a given component; β_i is the probability of deexcitation of vibrations as a result of collisions with the walls (this probability also allows for the average number of vibrational quanta transferred by diffusion).

Thus, the energy balance or rate equations (2.9) have the following structure. The first two terms in the brackets describe the excitation and deexcitation of vibrations by electron impact. The third term in the braces is responsible for the deexcitation of vibrations on the discharge-tube walls. The last two terms on the right-hand side of the equations in the system (2.9) describe the collisional relaxation processes, including exchange. These terms are obtained by summing the rate equations over all the vibrational levels of the oscillators (see preceding section) and thus, in contrast to [12], they allow for the role of the high levels.

Substituting from the system (2.10) into the system (2.9), we obtain a system of equations for x_i. The roots of this system give the vibrational temperatures and populations of the levels under consideration.

However, before solving this system of equations, we must have information on the parameters which occur in these equations, such as the exchange k_T and relaxation k_L rates or the quantities k^{ei}, $n_e(r)$, etc. All these depend on the probabilities of elementary processes and on macroscopic parameters of the electrical discharge. Since, in the final analysis, they determine the results of calculations, we must discuss them and select carefully the necessary data.

§ 3. Parameters of a Discharge Plasma and
Efficiencies of Elementary Processes

Relaxation Rates. Considerable success has recently been achieved in the determination of relaxation times of laser levels and of the transfer times of the vibrational energy. The most reliable results were obtained by the method of giant pulses suggested by Hocker and Kovacs [83] and developed further by Moore et al. [80]. Similar results were obtained by the method of time scanning the output radiation using short high-current pulses for the excitation of a plasma, as suggested and put into practice by Cheo [81].

Table 2 gives the rates of exchange k_T and deactivation k_L (sec^{-1} · Torr^{-1}) obtained by different workers at 300°K $(A^* + B \xrightarrow{k_T} A + B^*, \ A^* + B \xrightarrow{k_L} A + B + \Delta E)$.

For the sake of comparison, we shall now quote the rates of relaxation of the 01^10 level in a system of the deformation vibrations of the CO_2 molecule, which govern the rate of deactivation of the lower laser level in collisions with various particles (sec^{-1} · Torr^{-1}):

CO_2	N_2	CO	He
193 [80]	400 [80]	25 000 [81]	3400 [80]

TABLE 2. Rates of Vibrational Energy Exchange and of Vibrational Deactivation in Collisions

B	$A^* = CO_2\,(\nu_3)$		$A^* = CO$		$A^* = N_2$	
	k_T	k_L	k_T	k_L	k_T	k_L
CO_2		320 [84] 350 [80] 385 [81]	$8 \cdot 10^3$ [80, 84]	200 [84]	$2 \cdot 10^4$ [80, 84]	100 [84]
CO	$8.8 \cdot 10^3$ [84, 80]	193 [81] 200 [80]		~ 0 [88]	$8 \cdot 10^2$ [85]	~ 0 [88]
N_2	$2.4 \cdot 10^4$ [80, 84]	100 [84] 106 [80] 115 [81]	$> 8 \cdot 10^2$ [85]	140 [85]		~ 0 [86, 87]
He		85 [80] 0—50 [81]		0 [86]		~ 0 [88]

Laser mixtures usually contain a considerable amount of helium and, as shown in Chapter I, there are always CO molecules in the discharge. Therefore, in such mixtures the rate of relaxation of the lower laser level exceeds considerably the rate of relaxation of the upper level and we are justified in assuming that the gas (translational) temperature is equal to the vibrational temperatures T_1 and T_2 that we have introduced.

Since changes in the gas pressure, composition of the mixture, discharge current, dimensions of the discharge tube, and cooling conditions alter the gas temperature, the question of the temperature dependences of the probabilities of resonant exchange and relaxation as a result of collisions is important. These dependences were determined experimentally for $CO_2 - CO_2$ and $CO_2 - N_2$ collisions [84] and the results are plotted in Figs. 24 and 25.

According to [88], the deactivation of the CO vibrations is weak in the pure gas and in mixtures with many others (He, Ne, Ar, ...). However, the CO_2 molecules give rise to a finite rate of relaxation of the vibrational energy of CO. The most likely process is as follows. Ac-

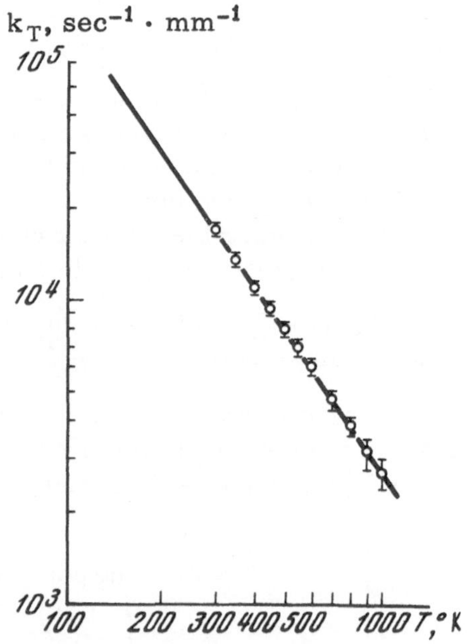

Fig. 24. Dependence of the rate constant k_T of the vibrational energy exchange between the N_2 and CO_2 molecules (antisymmetric vibrations) on the gas temperature [84].

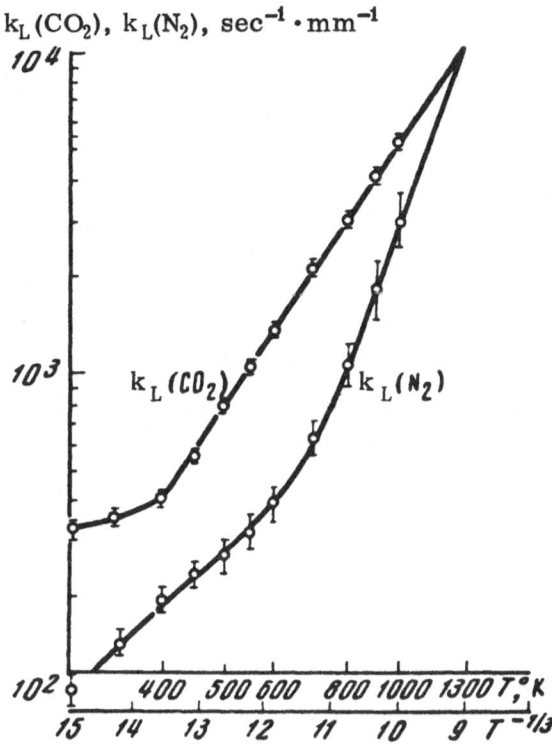

Fig. 25. Dependences of the rate of relaxation of the vibrational energy, stored in the antisymmetric vibrations of the CO_2 molecules and lost in CO_2-CO_2 and $CO-N_2$ collisions, on the gas temperature [84].

cording to [80], the probability of the process

$$CO_2(v=1) + CO_2(00^00) \rightarrow CO_2(11^10) + CO(v=0) \tag{2.11}$$

is $p = 1.9 \cdot 10^{-5}$, which corresponds to a rate constant $\sim 2 \cdot 10^2$ sec$^{-1} \cdot$ Torr^{-1}, whereas the rate of deactivation of the deformation (including symmetric) vibrations in CO_2-CO collisions is $\sim 2.5 \cdot 10^4$ sec$^{-1} \cdot$ Torr^{-1}. Therefore, we may assume that the converse process is unimportant and the transfer to the 11^10 level results in fast relaxation. The same relaxation process may account for the fact that the deactivation of the antisymmetric vibrations of CO_2 in collisions with CO is more effective than in collisions with N_2 (Table 2). This occurs via the intermediate exchange

$$CO_2(00^01) + CO(v=0) \rightarrow CO_2(00^00) + CO(v=1)$$

followed by the transfer of energy to the combination level 11^10, which is in better resonance with CO than with N_2.

Unfortunately, the temperature dependence of the rate of relaxation in CO_2-CO and CO_2-He collisions has not yet been determined experimentally. We shall therefore assume that $\ln p \propto T^{1/3}$, which follows from the Landau and Teller theory [73] (see § 1). Moreover, the temperature dependence of the rate of the $CO_2(\nu_3)-CO$ exchange of vibrational quanta is not known. However, in the latter case, we cannot extrapolate theoretically toward higher temperatures. The theory [68, 71] predicts $Q_{v,\ v-1}^{v',\ v'+1} \propto T$, whereas the experimental results [84] show that the rate of transfer decreases with rising temperature in the case of the $CO_2(\nu_3)-N_2$ exchange. We shall assume that the temperature dependence of the rates of the $CO-CO_2(\nu_3)$ and $CO-N_2$ exchange are of the same nature as those of the CO_2-N_2 process at room temperature except that they are 2.4 times smaller in the absolute sense. This assumption is justified because the results of calculations depend weakly on the rate of exchange provided it is much faster than the rate of relaxation, which is always true at moderately high gas temperatures

($\lesssim 700°K$). In cases of practical interest, the gas temperature is less. For example, when the rate of the CO_2-CO exchange is altered by a factor of 3, the calculated population of the upper laser level in the case of a CO_2-CO-N_2-He (0.4 : 0.6 : 3 : 6) mixture at 400°K changes by just 7%. The physical meaning of this result is self-evident: As long as the exchange process ensures a sufficiently rapid mixing of the vibrations, all of them relax at the same rate.

Allowance for non-Maxwellian Electron Velocity Distribution. Since molecular vibrations are excited by electron impact, it is essential to have more accurate data on the electron component of the discharge, i.e., on the electron density and distribution of the electron velocities. This information is available in the work of Novgorodov et al. [46]. A particularly important result can be found in the latter work: It is a reliable experimental determination of the nature of the electron velocity distribution function which is far from Maxwellian.

We shall now give the values of the cross sections (averaged over the electron velocities) for the excitation of various molecular vibration modes by electron impact:

i	$CO_2\,(\nu_3)$	N_2	CO	$CO_2\,(\nu_1)$	$CO_2\,(\nu_2)$
k^{ei}, cm^3/sec	10^{-8}	$1.2\cdot10^{-8}$	$2.4\cdot10^{-8}$	$3.6\cdot10^{-10}$	$1.6\cdot10^{-9}$

The quantities

$$k^{ei} < \langle v_e\sigma_i\rangle = \frac{\int f(v_e)\,\sigma_i\,(v_e)\,v_e\,dv_e}{\int f(v_e)\,dv_e}$$

were obtained by graphical integration using the electron velocity distribution functions $f(v_e)$ taken from [10, 46]. Information on the excitation cross sections $\sigma_i(v_e)$ of the symmetric and antisymmetric CO_2 vibrations, and of the N_2 and CO vibrations (i = 1, 3, 4, 5), was taken from [10, 89], whereas the semiempirical excitation cross section $\sigma_i(v_e)$ of the deformation vibrations of the CO_2 molecule (i = 2) was taken from [90].

As mentioned above, the values of $\langle v_e\sigma_i\rangle$ for the symmetric and deformation vibrations are small and cannot ensure a significant population of the relevant levels so that we can ignore the corresponding processes.

Since the electron excitation of the CO_2 molecule affects effectively only one quantum of antisymmetric vibrations $(\langle v\sigma\rangle_{00°2}/\langle v\sigma\rangle_{00°1} \approx 3\cdot10^{-2})$, whereas in the N_2 and CO molecules up to 8 quanta are excited [10, 89], it was assumed that $\alpha_3 = 1$ and $\alpha_4 = \alpha_5 = 3.5$, which represents a rough allowance for the relative efficiencies of the simultaneous excitation of different numbers of quanta.

Gas Temperature. Since the probabilities of relaxation processes depend strongly on the gas temperature, measurements of this temperature are of basic importance. Information on the gas temperatures in discharges was taken from [43]. Unfortunately, most of the results reported in that paper were obtained under nonoptimal conditions. Therefore, the lacking values were calculated using an approximate formula suggested in [91] for the determination of the gas temperature T_g averaged over the tube radius:

$$Q(T_g) = (\bar{T}_g - T_w)\,18.9\lambda_\Sigma(\bar{T}_g). \tag{2.12}$$

Here, $Q(T_g) = i_d E$ is the power dissipated in the gas per unit length of the discharge; T_w is the temperature of the discharge-tube walls; $\lambda_\Sigma(T_g)$ is the thermal conductivity of the gas mixture; i_d is the discharge current. The values of the electric field obtained in a wide range of currents, pressures, and compositions of the gas mixtures were taken from [46, 92].

The thermal conductivity of the gas mixture was calculated from the formula [93]

$$\lambda_\Sigma = \frac{\lambda_{CO_2}}{1+0.81\,\dfrac{p_{N_2}}{p_{CO_2}}+0.23\,\dfrac{p_{He}}{p_{CO_2}}} + \frac{\lambda_{N_2}}{1+1.4\,\dfrac{p_{CO_2}}{p_{N_2}}+0.34\,\dfrac{p_{He}}{p_{N_2}}} + \frac{\lambda_{He}}{1+3.4\,\dfrac{p_{CO_2}}{p_{He}}+2.7\,\dfrac{p_{N_2}}{p_{He}}}. \qquad (2.13)$$

The thermal conductivities of pure gases were taken from [94].

Following the theory of the positive column in electrical discharges in gases [44], it was assumed that the radial distribution of the electron density in a discharge tube was described by a zeroth-order Bessel function

$$n_e(r) = n_e(0)\,J_0\!\left(2.4\,\frac{r}{R}\right), \qquad (2.14)$$

where R is the discharge-tube radius.*

The same distribution was assumed for the excited molecules. It yielded a simple expression D_i/Λ^2 for the probability of diffusion decay of the excited states. The diffusion length was $\Lambda = R/2.4$. The concentrations of the excited molecules were assumed to be low and the diffusion coefficients in the gas mixture with specified partial pressures p_i were calculated from the formula [96]

$$\frac{1}{D_i} = \sum_{j=1,\,\ldots,\,i,\,\ldots,\,n} \frac{p_i}{D_{ij}},$$

where n is the number of components in a mixture.

The self-diffusion and diffusion coefficients of binary mixtures were taken from [94]. When the probabilities of the various processes and the values of the discharge parameters were selected, numerical calculations were made of the populations of the vibrational levels and the temperatures.

§ 4.   Solution of Relaxation Equations.
Results of Calculations

The substitution of the expressions (2.10) into the system (2.9) subject to Eq. (2.14) gives the following system of linearized equations if terms of the second order of smallness are ignored

$$\begin{aligned}
&(-k_T^{35}+2a_5k^{e5}n_e)\,x_3 + (2a_5k^{e5}n_e - k_T^{45})\,x_4 + \Big(0.62a_5k^{5e}n_e + \beta_5\frac{D_5}{\Lambda^2} + \\
&\qquad + 2a_5k^{e5}n_e + k_T^{35} + k_T^{45} + k_L^5\Big)\,x_5 = a_5k^{e5}n_e, \\[2mm]
&(-k_T^{34}+2a_4k^{e4}n_e)\,x_3 + \Big(0.62a_4k^{4e}n_e + \beta_4\frac{D_4}{\Lambda^2} + 2a_4k^{e4}n_e + k_T^{34} + k_T^{54} + \\
&\qquad + k_L^4\Big)\,x_4 + (2a_4k^{e4}n_e - k_T^{53})\,x_5 = a_4k^{e4}n_e, \\[2mm]
&\Big(a_3k^{3e}0.62 + k_T^{53} + k_T^{43} + \beta_3\frac{D_3}{\Lambda^2} + 2a_3k^{e3}n_e + k_L^3\Big)\,x_3 + (-k_T^{43} + 2a_3k^{e3}n_e)\,x_4 + \\
&\qquad + (2a_3k^{e3}n_e - k_T^{53})x_5 = a_3k^{e3}n_e.
\end{aligned} \qquad (2.15)$$

---

* The validity of the positive column theory is limited. Therefore, an experimental determination was made of the radial distribution [95], which confirmed the Bessel distribution. Recent and more accurate results obtained by Novgorodov (private communication) also confirmed this result.

The solution of the system (2.15) for x_i followed by the substitution of the results obtained in Eq. (2.10) gives the vibrational temperatures T_i and the populations of the vibrational levels. If this is followed by the calculation of the population of the lower level, which — in accordance with our assumptions — is governed by the gas temperature, we can finally calculate the population inversion in the investigated gas.

Figure 26 gives the results of a calculation of the population inversion in a mixture of 2 Torr CO_2 and p Torr N_2 without allowance for the dissociation of CO_2. Curve 2 in Fig. 26 is theoretical [12]; curve 1 gives the experimental values of the inversion deduced from the measured laser gain [99]. Curves A and B are the results of our calculation obtained for different dependences of the gas temperature on the nitrogen pressure (lines II and I, respectively: line I was calculated using the field intensities given in [46] and line II was plotted for comparison). We can see that our results describe well the experimental dependences. However, full agreement cannot be expected because changes in the gas composition under the influence of a discharge, which do occur in reality, have been ignored.

It is worth noting (see Fig. 26) that the temperature has a strong influence on the results (compare curves I and II). In general, the dependence of the inversion density on the gas pressure is due to several factors such as changes in the concentrations of the molecules and in the electron density, as well as changes in the degree of dissociation and the electron velocity distribution function. However, the greatest contribution is made by the change in temperature, mainly because of the strong temperature dependences of the relaxation rates. For this reason, a quantitative agreement between the calculated and experimental results can be expected only if the dependence of the gas temperature parameters and the temperature dependences of the relaxation rates are known in detail.

The general nature of the dependences in Fig. 26 can be explained qualitatively as follows. Since the pressure dependences of the electron density and electron velocity distribution function are weak, an increase in the concentration of nitrogen molecules is accompanied by a parallel increase in the rate of vibrational pumping per unit volume. An increase in the molec-

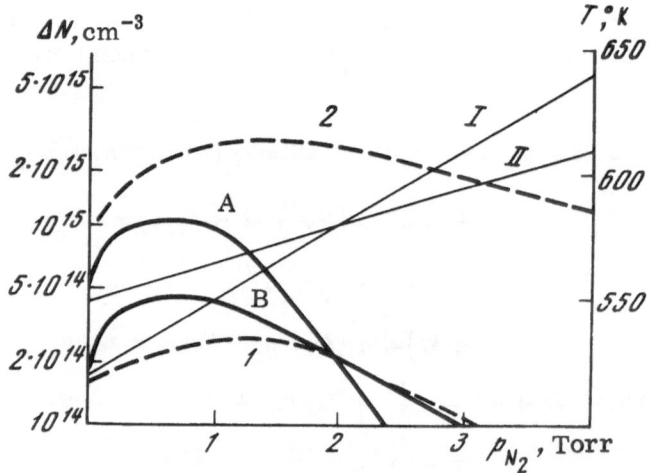

Fig. 26. Population inversion in a mixture of 2 Torr CO_2 + p Torr N_2 in a water-cooled tube of 34 mm diameter; i_d = 30 mA.

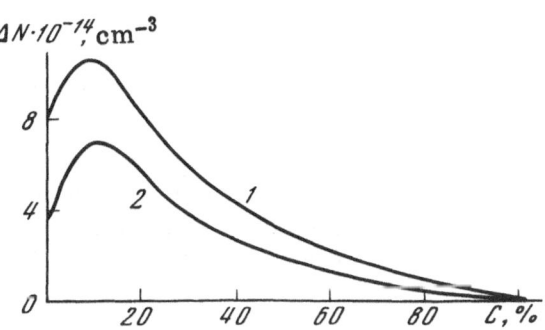

Fig. 27. Dependence of the population inversion density on the degree of dissociation of CO_2 in two mixtures: 1) CO_2-N_2-He (1:3:6), p_Σ = 6.9 Torr, T = 500°K; 2) CO_2-N_2 (1:1), p_Σ = 2 Torr, T = 400°K. The diameter of a water-cooled tube was 22 mm.

ular concentration needed to maintain a constant current requires a higher electric field and it follows from Eq. (2.12) that this is accompanied by an increase in the gas temperature. The rate of relaxation of the upper laser level and the population of the lower laser level increase in parallel. The competition between these two factors is responsible for the existence of an optimal pressure.

Figure 27 gives the results of calculation of a population inversion in CO_2-N_2-He and CO_2-N_2 mixtures as a function of the degree of dissociation of carbon dioxide. It is clear from this figure that the inversion depends strongly on the degree of dissociation of CO_2. It is important to note that this dependence is nonmonotonic and has a maximum for low degrees of conversion of CO_2 into CO. This makes it possible to optimize the operation of the CO_2 laser by varying the rate of flow of the gas or by adding gases which affect the rate of dissociation. Estimates indicate that a change in the degree of dissociation of CO_2 has only a slight influence on the gas temperature. Therefore, the existence of a maximum should be attributed to other factors. The initial increase in the degree of dissociation of the CO_2 molecules is accompanied by a fall in their concentration but it also increases the population inversion because of the effective excitation of the CO molecular vibrations by electron impact followed by the transfer of energy to the antisymmetric vibrations of the CO_2 molecules. An undesirable factor is the increase in the rate of relaxation of the vibrational energy with increasing CO content in accordance with the mechanism suggested in [5, 9, 11]. These factors are responsible for the presence of maxima in Fig. 27.

Since the times for the establishment of an equilibrium degree of dissociation are comparable with the time spent by the continuously flowing gas in the discharge (Chap. I), the gas composition must be strongly inhomogeneous along the discharge tube. If we use the kinetic curves (see Chap. I) and the dependences in Fig. 27, we can find the distributions of the population inversion along the length of a laser tube. These distributions are plotted in Fig. 28 for a 60-cm long tube and different flow velocities. Figure 29 shows the dependence of the inversion on the time spent by the laser mixture in the discharge. Thus, it is clear that the values of the population inversion deduced from the low-signal gain give a value averaged over the length

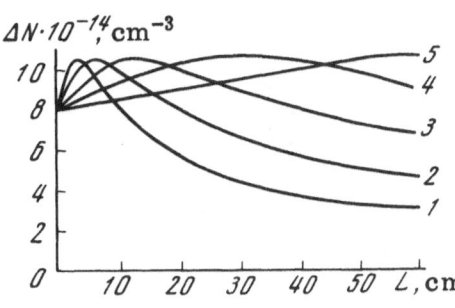

Fig. 28. Distributions of the inverted population along the discharge tube in a continuous-flow system calculated for different flow velocities (m/sec): 1) 0.3; 2) 0.6; 3) 1.2; 4) 2.4; 5) 4.8. The diameter of a water-cooled tube was 22 mm.

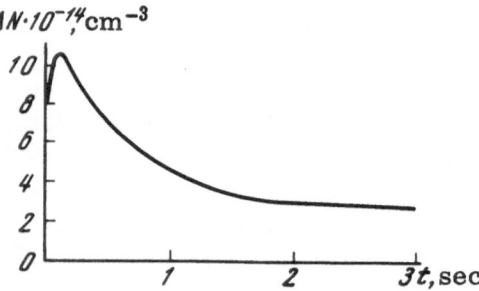

Fig. 29. Dependence of the population inversion density on the time spent by a CO_2-N_2-He (1 : 3 : 6) mixture in a discharge at p = 6.9 Torr, i_d = 30 mA. The tube diameter was 22 mm.

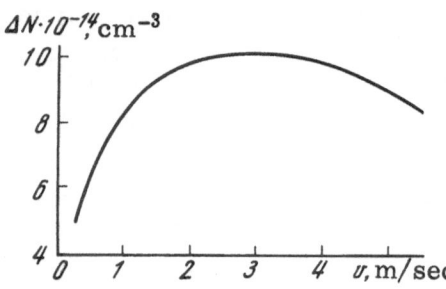

Fig. 30. Dependence of the population inversion density, averaged over the length (60 cm) of a discharge tube of 22 mm diameter, on the velocity of flow of a CO_2-N_2-He (1 : 3 : 6) mixture (p_Σ = 6.9 Torr, i_d = 30 mA).

of the discharge tube

$$\overline{\Delta N_i} = \frac{1}{L} \int_0^L \Delta N(l)\, dl. \tag{2.16}$$

The dependence of the inversion on the linear flow velocity in a 60-cm long tube is plotted in Fig. 30. The rapid rise of the inversion at low flow velocities slows down eventually and shows saturation at moderate velocities. This is followed by a fall in inversion at higher velocities. Clearly, in the case of longer tubes, this fall occurs at even higher flow velocities.

Let us also consider discharges in pure CO_2 when no dissociation takes place. In this case, we cannot assume that the levels of the symmetric and deformation vibrations have populations in equilibrium with the gas temperature because the rate of deactivation of these levels as a result of CO_2-CO_2 collisions is low. However, even if this assumption is made, i.e., if an estimate is obtained of the minimum likely population of the lower laser level, calculations show that there should be no inversion because the gas temperature of pure CO_2 is too high. Figure 31 shows our calculated dependences of the populations of the upper and lower laser levels on the pressure of CO_2.

Fig. 31. Pressure dependences of the population of the 00^01 level and of the equilibrium population of the 10^00 level of the CO_2 molecule in pure CO_2 gas. The tube diameter was 34 mm, i_d = 30 mA. The values of the gas temperatures were taken from [99].

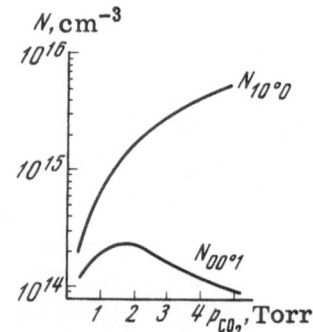

Role of Individual Components. An analysis of the results of calculations based on the latest experimentally obtained probabilities of elementary processes of excitation and deexcitation of molecular vibrations and on the values of laser discharge parameters gives the fullest insight into the role of the individual components of a gas mixture in the establishment of a population inversion in a carbon dioxide laser.

Carbon dioxide is the major constituent of the active gas. The laser transitions occur between vibrational levels of the carbon dioxide molecule. The antisymmetric vibrations, including the upper laser level, are pumped fairly strongly by electron impact. Nevertheless, population inversion does not occur in pure CO_2. This is because the gas temperature is too high and, consequently, the rate of relaxation of the upper level is too fast and the population of the lower level is too high. Moreover, CO_2 dissociates rapidly.

Vibrations of the nitrogen molecules are excited effectively by electron impact and are in resonance with the antisymmetric vibrations of CO_2. In spite of the fact that the rate of excitation of these nitrogen vibrations is only slightly higher than that of CO_2, the addition of nitrogen gives rise to population inversion. This is due to the following reasons:

a) the rate of relaxation of the energy of the antisymmetric CO_2 vibrations in $CO_2 - N_2$ collisions is 2-3 times less than in $CO_2 - CO_2$ collisions;

b) the replacement of some of the CO_2 molecules with N_2 reduces somewhat the gas temperature;

c) when vibrations are excited, it is very likely that several (up to eight) quanta are excited simultaneously so that, in the case of similar values of $\langle v\sigma \rangle$ for N_2 and CO_2, the energy stored in the vibrational modes of nitrogen is greater.

Carbon monoxide is pumped very effectively by electron impact and, as in the case of N_2, up to eight vibrational quanta are excited simultaneously. This, in combination with the rapid exchange of vibrational quanta in $CO_2 - CO$ collisions, increases the density of population inversion for low concentrations of CO. However, when the dissociation of carbon dioxides gas ($CO_2 \rightleftarrows CO + O$) proceeds too far, it has a negative effect on the inversion primarily because of the considerable rate of relaxation of the vibrational energy in $CO - CO_2$ collisions in accordance with the mechanism

$$CO(1) + CO_2(00^00) \rightarrow CO(0) + CO_2(11^10),$$
$$CO_2(11^10) + M \rightarrow CO_2(00^00) + M + \Delta E,$$

and because of a reduction in the amount of CO_2. The lower laser level is also deactivated very rapidly.

Helium reduces the gas temperature and increases somewhat the electron density. However, a large amount of helium (over 90%) in a gas mixture has a negative effect on the inversion because of the finite rate of relaxation of the energy of the antisymmetric CO_2 vibrations in $CO_2 - He$ collisions. Once again, the lower laser level is deactivated rapidly.

CHAPTER III

EXPERIMENTAL DETERMINATION OF LASER LEVEL POPULATIONS. VIBRATIONAL TEMPERATURES

The results of calculations (carried out in the preceding chapter) of the population inversion are in satisfactory agreement with the experimental results deduced from measurements of the laser gain [92, 97]. However, it is clear from Eqs. (2.9) and (2.10) that the direct results of such calculations are the vibrational temperatures of molecules in different normal modes and of the absolute populations of the corresponding vibrational levels. The population inversion is then calculated as the difference between the populations of the upper and lower laser levels. Therefore, in a more detailed check of the population inversion mechanisms predicted theoretically it would be desirable to use experimental data on the populations of individual vibrational levels.

Soviet [66, 98] and overseas [40, 99] investigators have suggested a method for the separate determination of the populations of the upper and lower laser levels. In this method the gain is measured for individual vibration-rotational transitions in various vibrational branches of the CO_2 molecule. Population inversions found this way are in agreement with the direct determinations of the inversion and are also close to the results of our calculations. However, the populations of the lower level 10^00 found in this way are considerably greater than the equilibrium values at the gas temperature, which cannot be explained theoretically because the rate of relaxation (depopulation) of this level is high (at least in the presence of considerable amounts of helium in the gas mixture). Moreover, the method itself is based on the assumption that the translational motion of a gas and the rotation of the CO_2 molecule are in thermodynamic equilibrium. This assumption is fully justified (see § 1 in preceding chapter). In fact, experimental measurements of the gain of individual rotational lines of laser transitions show that there is a Boltzmann distribution of the molecules between the rotational levels. However, as established in [40], the temperature calculated in this way is very different from the temperature deduced from the distribution of intensities in the rotational structure of the spontaneous luminescence in the region of the second positive system of nitrogen. For example, the temperature found experimentally (by the laser method) is 340°K [40], whereas the distribution of the N_2 molecules (state $C^3\Pi_u$) over the vibrational levels corresponds to 535°K. The cause of this divergence is not clear. It would be desirable to check the method described above by an independent technique because measurements of the laser level populations are essential for the understanding of the mechanism responsible for the CO_2 laser action. In the present chapter we shall consider qualitative and quantitative dependences of the populations of the upper laser level on various discharge parameters and we shall do this on the basis of the spontaneous radiation (luminescence) intensities.

§ 1. Method for Measuring Vibrational Temperatures and Populations of Vibrational Levels on the Basis of Visible and Ultraviolet Spontaneous Radiation

Statement of the Problem

In investigations of the CO_2 laser action it is usual to confine attention to the vibrational levels belonging to the ground electronic state. This is in agreement with the generally accepted view that highly excited electronic states play a negligible role in the establishment of a population inversion between the vibrational levels of the ground state (the exception to this rule is the first Patel hypothesis, which has not been confirmed—see Introduction).

However, there is a coupling between the electronic ground and the excited states and it is manifested by modulation of the visible side bands when the Q factor of the resonator laser is switched. Such radiation appears as a result of transitions between different electronic states of the CO_2, CO, and N_2 molecules [100, 101]. This provides a basis for the determination of the populations of the vibrational levels of the molecules in the ground electronic states from the visible and ultraviolet radiation.

The proposed method is based on the measurement of the relative intensities of the vibronic bands [102].

The method of measuring the relative intensities is used widely in spectroscopy. A well-known method for the determination of the temperatures of molecular gases is based on the distribution of intensities in the rotational structure of vibronic bands. It is shown in Chapter II that the thermalization of the rotational states occurs in a time of the order of the molecular mean free time ($\sim 10^{-7}$ sec at a pressure of 1 Torr) and, therefore, the distribution of molecules between the rotational levels in the ground electronic state is clearly of the Boltzmann type with a temperature equal to the temperature of translational degrees of freedom. A completely different situation is encountered in the case of highly excited electronic states. It is frequently found that the radiative lifetimes of these states are comparable to or even shorter than the molecular mean free times, i.e., 10^{-7}-10^{-8} sec, so that the thermalization does not occur in the available time. However, it is shown in [103] that in these cases the distribution of molecules between the vibrational levels is close to the Boltzmann form but with the translational temperature, i.e., the ground state distribution is "copied." Physically, this is due to the fact that the molecular momentum is only slightly affected by a collision between a molecule and an electron.

Clearly, this does not mean that the copying process applies also to the distributions between the vibrational levels in the ground and excited electronic states. In spite of that, frequently no distinction is made between the vibrational temperatures of the ground and electronically excited states (see, for example, [104-107]). Thus, the method of measuring the distributions of molecules between the vibrational levels in the ground electronic state on the basis of the intensities of bands which begin from the vibrational levels of electronically excited states should be used to establish first the relationship between these distributions.

Method

The good agreement between the energies of the vibrational levels of the antisymmetric vibrations of the CO_2 molecule and of the vibrations of the CO and N_2 molecules in their gound electronic states demonstrates a strong coupling between these vibrations. For example (see Table 2), the rate k_T of the exchange of quanta in CO_2-N_2 collisions at 300°K is $2 \cdot 10^4$ sec$^{-1} \cdot$ Torr^{-1}, whereas the rate of conversion of the vibrational to the translational energy by the same collisions is $\sim 10^2$ sec$^{-1} \cdot$ Torr^{-1}. Therefore, it is sufficient to know the populations of the vibrational levels of any one of these molecules to find the populations of all three. This method can be used to find the populations of the vibrational levels of the N_2 molecules in the ground state $X^1\Sigma_g^+$ by measuring the intensities of the bands in the 2^+ (second positive) system of nitrogen. This system appears as a result of transitions between the electronic states $C^3\Pi_u$ and $B^3\Pi_g$ (Fig. 32). We can show (this point is discussed later) that in the discharge plasma under discussion here (electron densities n $\sim 10^9$-10^{10} cm^{-3}, average electron energy $\bar{\varepsilon} = 2$-3 eV), the upper electronic state $C^3\Pi_u$ is populated by direct electron impact from the gound electronic state. The cross section of this process is fairly large and can reach values of 10^{-16} cm^2 [108, 109].

On the other hand, the vibrational levels of the $C^3\Pi_u$ state are depopulated exclusively by the radiative process because the typical lifetime of this process is 10^{-8} sec [110], which is much shorter than the vibrational and rotational thermalization times, and very much shorter than the times needed for the diffusion to the walls under conditions typical of carbon dioxide lasers.

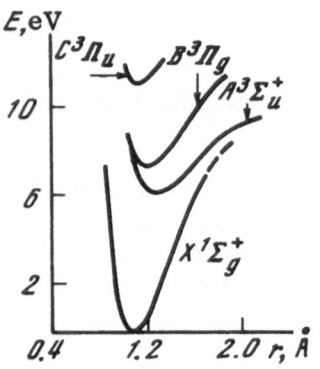

Fig. 32. Some electronic states of the N_2 molecule.

Let us assume that the distribution of the N_2 molecules between the vibrational levels v_X'' in the ground state is described by the distribution function $f(v_X'')$, i.e.,

$$N_{v_X''} = N_0 f(v_X'). \tag{3.1}$$

Then, under the assumptions made above, the steady-state population of the vibrational levels v_C' of the state $C^3\Pi_u$ can be described by

$$N_{v_C'} = \frac{n_e N_0}{A_{v_C' v_B''}} \sum_{v_X''} \langle v\sigma \rangle_{v_C' v_X''} f(v_X''), \tag{3.2}$$

where N_0 is the number of molecules in the $X^1\Sigma_g^+$ state with $v_X'' = 0$; $\langle v\sigma \rangle_{v_C' v_X''}$ is the average excitation cross section of the vibrational levels v_C' of the state $C^3\Pi_u$ from the state $X^1\Sigma_g^+$; n_e is the electron density; $A_{v_C' v_B''}$ is the probability of emission of the 2^+ system bands; v_B'' are the vibrational $B^3\Pi_g$.

Unfortunately, the published data on the probabilities of the excitation of the vibrational levels of the $C^3\Pi_u$ state from the ground state are incomplete. However, since the electron-impact excitation is a fast process which is completed in a time much shorter than the molecular vibration period, these probabilities can be regarded [74] as proportional to the Franck–Condon factors of the $C^3\Pi_u - X^1\Sigma_g^+$ Tanaka system. Then, Eq. (3.2) becomes

$$N_{v_C'} = C \sum_{v_X''} q_{v_C' v_X''} f(v_X''), \tag{3.3}$$

where C is a constant independent of v_X''. Bearing in mind that the vibrational level populations decrease rapidly with increasing v, we can simplify Eqs. (3.2) and (3.3) by retaining only the first five terms. If we know the relative populations of five levels of the $C^3\Pi_u$ state we can then find the relative populations of the vibrational levels of the gound state by solving the system

$$N_{v_C'} = \alpha \sum_{v_X''=0}^{4} q_{v_C' v_X''} N_{v_X''} \tag{3.4}$$

for $N_{v_X''}$ with $v_C' = 0, 1, 2, 3, 4$; α is a constant coefficient.

The relative populations of the vibrational levels of the $C^3\Pi_u$ state can be determined experimentally in the usual way from the relative intensities of the 2^+ system of N_2 ($C^3\Pi_u - B^3\Pi_g$ transition).

It follows from the molecular spectroscopy theory that the intensity of a vibronic band is

$$I_{v'_C v''_B} = C' S_e N_{v'_C} \nu^4_{v'_C v''_B} q_{v'_C v''_B}, \tag{3.5}$$

where S_e is the electronic oscillator strength, $q_{v'_C v''_B}$ are the Franck–Condon factors, $\nu_{v'_C v''_B}$ is the transition frequency, and C' is a constant independent of the quantum numbers v'_C and v''_B. Taking logarithms of Eq. (3.5), we obtain

$$\ln N_{v'_C} = \ln C' + \ln \frac{I_{v'_C v''_B}}{q_{v'_C v''_B} \nu^4_{v'_C v''_B}}. \tag{3.6}$$

It is assumed here that S_e is independent of v'_C and v''_B [111].

Thus, measurements of the relative intensities of the vibronic bands beginning from a sequence of the v'_C levels can yield the relative populations of the v''_X levels and the vibrational temperatures of the N_2 molecules in the ground state (and at the same time the absolute populations of the v''_X levels because — under the conditions considered here — the majority of the N_2 molecules are in the ground electronic state).

We shall first consider the conditions of applicability of the method and then describe the measurements.

Justification of the Method

We can justify the determination of the vibrational temperatures of the N_2, CO, and CO_2 molecules in the ground state (antisymmetric vibrations) by the method described above but we must confirm, first of all, the basic assumption of the direct electron excitation of the $C^3\Pi_u$ state and, secondly, the assumption of a strong collisional coupling between the vibrational levels of these molecules.

Excitation of the $C^3\Pi_u$ State. The assumption of the direct electron excitation of the $C^3\Pi_u$ state can be checked by analyzing the discharge conditions in a tube of 20 mm diameter carrying a discharge current of 0–50 mA and containing a $CO_2 - N_2 - He$ mixture of gases at a total pressure up to 10 Torr.

The presence of strongly populated metastable states is essential for multistage (cascade) excitation. In our case there is the $A^3\Sigma_u^+$ state with a radiative lifetime $\tau \sim 1$ sec [112]. Let us estimate the upper limit of the population of this state under our conditions. We shall assume that the state is depopulated only as a result of diffusion to the tube walls and that it is populated both by direct electron impact and by radiative decay of the $C^3\Pi_u$ and $B^3\Pi_g$ levels (see Fig. 32). We shall take the necessary cross sections from [108] and the electron velocity distribution from [46]. It follows from these estimates that the steady-state population of the $A^3\Sigma_u^+$ state does not exceed 10^{12}–10^{13} cm^{-3} and that the multistage excitation is important when the cross section of the process $N_2(A^3\Sigma_u^+) + e \rightarrow N_2(C^3\Pi_u) + e$ is $\sim 10^{-13}$ cm^2, which is unlikely. This is largely due to the fact that, as established in [46], the electron energy distribution function for discharges of this type departs strongly from the Maxwellian form. A characteristic feature which distinguishes the distribution from the Maxwellian function is a strong cutoff of the high-energy part. Consequently, the rate of excitation of the $C^3\Pi_u$ state from the $A^3\Sigma_u^+$ state is low (the $C^3\Pi_u$ state is located about 5 eV above $A^3\Sigma_u^+$).

The processes of multistage excitation manifest themselves much earlier in the population of the $B^3\Pi_g$ state which lies only 1.5 eV above $A^3\Sigma_u^+$.

The hypothesis of direct electron excitation of the $C^3\Pi_u$ state was checked experimentally by measuring the dependences of the intensities of the 2^+ band system on the discharge current

(0-30 mA) at different pressures of N_2 in mixtures with CO_2 and He. The linear nature of these dependences demonstrated direct excitation.

These conclusions are in agreement with those reached in [113] that in the case of a glow discharge in nitrogen the multistage excitation of the electron states is unimportant. The same paper [113] reports also the dependences of the intensities of the bands in the first and second positive systems of nitrogen on the discharge current. Measurements were carried out in a tube of 7 mm diameter at nitrogen pressures of 0.2-1.5 Torr. It was found that the dependences were linear throughout the investigated range of currents (5-100 mA) not only in the second but also in the first positive system.

In general, the linear nature of the dependences of the band intensities on the current does not exclude completely the possibility of multistage excitation of the states from which these bands begin. Such excitation may, in principle, occur if the population of the intermediate state (or several intermediate states) is independent of the discharge current. In this case the population saturates because of collisions of the second kind with electrons. Although the situation is different in our case (the electron densities are far too low), nevertheless, it would be useful to know the absolute populations of the possible intermediate (in our case $A^3\Sigma_u^+$) states. The upper limit of the population of the $A^3\Sigma^+$ state can be estimated using the results given in [114], which reports measurements of the absorption corresponding to the $A^3\Sigma_u^+ - B^3\Pi_g$ transition. These measurements were made by the "line into line" absorption method, whose sensitivity was proportional to the probability of the transition in question. No absorption was observed under all the investigated discharge conditions (a tube of 7 mm diameter was used, the nitrogen pressure was 0.2 Torr, and the discharge current was 0-150 mA), which indicated that the density of the $A^3\Sigma_u^+$ states was less than 10^{12} cm^{-3}, i.e., it was in agreement with the results of our theoretical estimates. A simple analysis shows that the population of the $C^3\Pi_u$ state by a multistage process is also unimportant.

It follows from the above considerations that the $C^3\Pi_u$ state is excited by direct electron impact.

<u>Spontaneous Infrared Radiation Emitted by CO_2 Laser Discharges.</u> A strong collisional coupling between the vibrational levels of the CO_2, CO, and N_2 molecules was checked by measuring the intensities of the spontaneous infrared radiation emitted by the CO_2 and CO molecules in discharges [115]. Use was made of an MDR-2 high-luminosity monochromator and of a liquid-nitrogen-cooled InSb detector.

These measurements showed that the spontaneous radiation (luminescence) spectrum of CO_2 had a band corresponding to antisymmetric vibrations of CO_2 with a maximum at about $4.6\ \mu$, shifted by $0.3\ \mu$ relative to the absorption band ($4.3\ \mu$). A similar shift was reported earlier in [116, 117] and attributed to the efficient excitation of high vibrational levels by electron impact in the discharge. In view of the anharmonicity of the vibrational levels, the transitions between high levels could give rise to radiation of longer wavelengths than the $00^01 - 00^00$ transition. However, this explanation is unsatisfactory because it is established in [89] that the direct electron excitation of the antisymmetric vibrations with $v_3 > 1$ is not very effective. Therefore, it remains to assume that the presence of strongly populated high vibrational levels is due to the exchange of quanta within a given vibration mode and it can occur simply because of electron impact excitation of the first vibrational quantum.

This applies also to the spontaneous radiation spectrum of the CO molecules, which exhibits a similar shift. Measurements were made of the radiation emitted by the CO_2 and CO molecules in the presence of N_2. The addition of nitrogen to continuous-flow and sealed systems increased greatly the radiation intensity, which indicated a strong coupling between the vibrations of these molecules.

A comparison of the intensities of the infrared radiation emitted from continuous-flow and sealed laser systems shows that the intensities of the radiation emitted by CO_2 and CO are higher in the sealed case and this is observed both for pure CO_2 as well as for $CO_2 - N_2$ and $CO_2 - N_2 - He$ mixtures. Since the concentration of the CO molecules in a sealed system is higher than in a continuous-flow laser, the increase in the intensity may be attributed to the effective electron pumping of the vibrational levels of CO and the subsequent resonant exchange of energy between the CO, N_2, and CO_2 molecules as a result of collisions. The addition of H_2 and H_2O reduces strongly the infrared radiation emitted by CO_2 and CO because the probability of the depopulation increases. The addition of helium has little effect on the radiation intensity.

§ 2. Experimental Method

A glow discharge in the investigated gas mixtures was ignited in a tube of molybdenum glass whose internal diameter was 20 mm. The discharge gap was 460 mm. A static high voltage was applied to Al electrodes. The power supply used could provide voltages up to 10 kV and currents up to 50 mA. The tube was filled with N_2, CO_2, and He gases in various ratios. We used N_2 and CO_2 of technical grade. Most of the experiments were carried out under continuous flow conditions at a velocity of ~ 1 m/sec. The apparatus also provided facilities for sealing off the discharge system.

The discharge tube was placed in front of the entry slit of a DFS-8 spectrograph which received radiation from the tube end covered with a quartz window. This spectrograph was used in photographic recording of the spectra. However, in the present study it was more convenient to employ photoelectric recording. Therefore, an exit slit was placed in the plane where a photographic plate would have normally been located. The DFS-8 spectrograph was an instrument of high dispersion (6 Å/mm for a diffraction grating with 600 lines/mm). However, we did not have to resolve the rotational structure of the band (as pointed out above). Therefore, measures were taken to ensure that the recorded spectrum had only a weak rotational structure which could easily be dealt with by planimetry.

We used an FÉU-18A photomultiplier with a Uviol window. Since the measurements were carried out in a fairly wide range of wavelengths, we plotted the spectral sensitivity of the photomultiplier. This was done in the usual way (see Appendix II). The photomultiplier signal was amplified with a narrow-band amplifier and passed to a synchronous detector; this allowed us to eliminate largely the photomultiplier noise and errors due to the instability of the discharge supply source. The spectrum was recorded on the chart of an ÉPP-09M3 potentiometer.

§ 3. Results and Discussion

It follows from Eqs. (3.2)-(3.4) that the distribution of molecules between vibrational levels in an electronically excited state depends on the distribution between vibrational levels in the ground electronic state but — in general — these distributions are not identical.

We may assume, for example, that there is a Boltzmann distribution of the N_2 molecules in the ground electronic state, i.e., $f(v''_X) = \exp\{-E_{v''_X}/kT_{v''_X}\}$, where $E_{v''_X}$ is the energy of the vibrational levels v''_X; $T_{v''_X}$ is the vibrational temperature of the N_2 molecules in the $X^1\Sigma_g^+$ state; we shall now determination the distribution in the $C^3\Pi_u$ state. This can be done using Eq. (3.3). The results of our calculations are plotted in Fig. 33. This figure gives the dependences of the populations of the v'_C levels of the $C^3\Pi_u$ state on the energies of the corresponding vibrational terms (the Franck–Condon factors were taken from [118] and the values of the energy were measured from the $v''_B = 0$ level of the $B^3\Pi_g$ state). The different curves in this figure correspond to different vibrational temperatures in the ground electronic state $X^1\Sigma_g^+$. We can see that the distribution between the v'_C levels is not of the Boltzmann type (in the Boltzmann

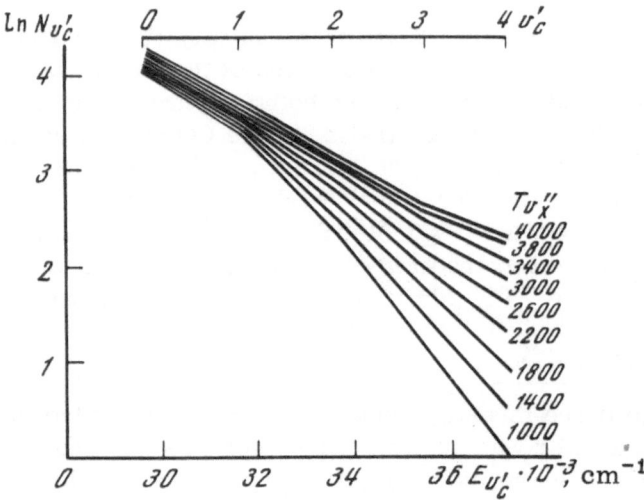

Fig. 33. Calculated distributions of the N_2 molecules between the vibrational levels of the $C^3\Pi_u$ state plotted for different values of the vibration temperatures $T_{v''_X}$ of the ground state $X^1\Sigma_g^+$.

case the dependences of $\ln N_{v'_C}$ and $E_{v'_C}$ would have been linear for all values of the vibrational temperature $T_{v''_X}$).

Clearly, the results obtained can be used also to solve the converse problem when the distribution of nitrogen molecules between the v'_C levels is known and found to be identical with one of the calculated distributions. This identity would, first of all, indicate a Boltzmann distribution of these molecules between the v''_X vibrational levels and, secondly, would allow us to determine the vibrational temperature $T_{v''_X}$, corresponding to this distribution. Such an identity can have only one meaning because the system of linear equations (3.4) has only one solution.

The relative populations of the vibrational levels of the $C^3\Pi_u$ state can be determined experimentally using Eq. (3.6) and measuring the relative intensities of the second positive system of nitrogen. Such measurements were carried out for N_2, CO_2-N_2 (1:2), CO_2-N_2 (2:1), CO_2-N_2 (9:1), and CO_2-N_2-He (1:3:8) at different currents and pressures. The series of bands with $\Delta v = -2$ and $U = -3$ were determined. In the calculations we used the Franck–Condon factors for the 2^+ system of nitrogen taken from [118].

The distributions of the populations obtained experimentally for N_2 at p = 1.37 Torr and for a CO_2-N_2 (1:2) mixture at p = 2.1 Torr are plotted in Fig. 34 (continuous curves). In

Fig. 34. Theoretical and experimental distributions of the N_2 molecules between the vibrational levels of the $C^3\Pi_u$ state in pure N_2 and in a CO_2-N_2 (1:2) mixture: 1) $T_{v''_X} = 4000°K$; 2) $T_{v''_X} = 1000°K$.

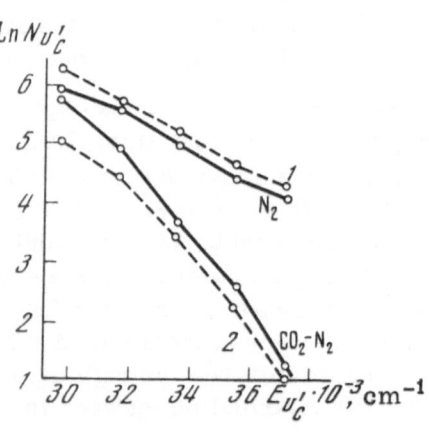

both cases the discharge current was $i_d = 20$ mA. The dashed curves were the results of calculations plotted in Fig. 33. In this case we selected the theoretical curves which fitted best the experimental dependences. We thus found that the experimental curve for nitrogen corresponded to a vibrational temperature of the $X^1\Sigma_g^+$ state, which was 4000°K, whereas the temperature of a mixture of nitrogen with carbon dioxide was about 1000°K. Similar comparisons could also be made for the remaining investigated cases. It should be noted that all the experimental distributions of the populations between the vibrational levels of the $C^3\Pi_u$ state (we obtained more than 50 such distributions) had characteristic kinks corresponding to kinks in the theoretical dependences. This confirmed the hypothesis that the distribution of the investigated molecules between the vibrational levels was of the Boltzmann type (see Chap. II).

As mentioned earlier, the relative populations of the levels in the $X^1\Sigma_g^+$ state and, consequently, the vibrational temperatures $T_{v_X''}$ could, in principle, be determined from the distributions of the populations in the $C^3\Pi_u$ state by solving the system (3.4). However, we encountered the following difficulties in this procedure. An analysis of the system (3.4) showed it to be very sensitive to the experimental error. In some cases the formal solution of the system (3.4) gave negative populations for the higher levels. The sensitivity was due to the following reason. The distribution of molecules between vibrational levels is described by a function $f(v)$ which decreases rapidly with a rise in v. Therefore, the experimental error in the determination of the populations of the lower levels may be greater than the absolute populations of the higher levels. This leads to misunderstandings. Clearly, the approach to the analysis of the experimental results should be modified. For example, we can determine the vibrational temperature of the ground state by selecting the experimental curves which fit best the calculated dependences, as is done in Fig. 34.

However, we used a different method. It is clear from Fig. 33 and from the experimental dependences of $\ln N_{v_C'}$ on v_C' that although there is no Boltzmann distribution in the $C^3\Pi_u$ state, the kinks in the dependences are not very sharp. If these kinks are rectified by, for example, the least-squares method, we can introduce the concept of $T_{v_C'}$, bearing in mind that it is purely nominal. Then, we can determine the temperature in the ground state by plotting an auxiliary dependence of the temperature $T_{v_X''}$ on the temperature $T_{v_C'}$. This graph is shown in Fig. 35. The results analysed in this way are plotted in Figs. 36 and 37.

The dependences of the vibrational temperatures on the pressure in various gas mixtures are plotted in Fig. 36. We can see that the vibrational temperature of pure nitrogen is much higher than the temperature of mixtures of N_2 with CO_2. This can be explained quite naturally by the efficiency of relaxation processes of molecules of different kinds, discussed in detail in Chapter II. It is clear from Table 2 that there is practically no relaxation of the vibrational energy in the $N_2^* - N_2$ collisions. Moreover, the nitrogen molecule is symmetric and its infrared dipole radiation is forbidden. Thus, in pure nitrogen the vibrational energy relaxes only by collisions of the second kind with electrons and by diffusion of molecules to the wall tubes. When the pressure is increased, the importance of the latter process decreases and the vibrational temperature should rise, as observed experimentally. In the case of $N_2 - CO_2$ mixtures the N_2 molecules can easily exchange energy with the antisymmetric vibrations of the CO_2 molecules and the energy of the latter molecules relaxes more rapidly (see Table 2) in $CO_2^* - CO_2$, $CO^* - N_2$, $CO_2^* - CO$, and $CO_2^* - He$ collisions. Therefore, we find that the vibrational temperatures in mixtures with CO_2 are low, as confirmed experimentally (lower curves in Fig. 36).

The dependences of the vibrational temperature on the current were determined in the range 10-30 mA. In this range the temperature increased (Fig. 37), which was clearly due to an increase in the density of electrons and a parallel increase in the rate of excitation of the vibrations.

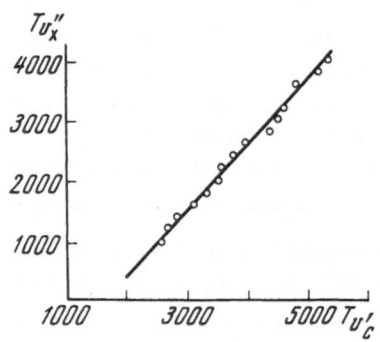

Fig. 35. Dependence of the vibrational tempera-ature of the ground state $T_{v''_X}$ on the vibrational temperature of the state $C^3\Pi_u$ in N_2.

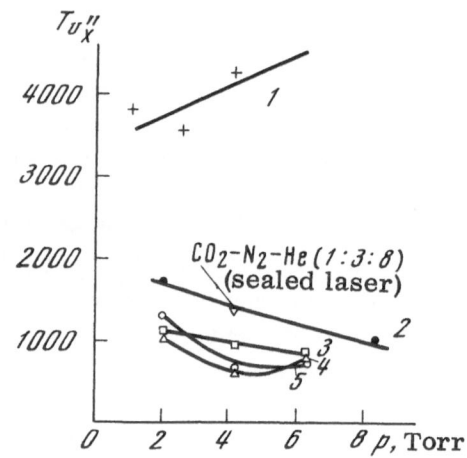

Fig. 36. Dependences of the vibrational temperatures $T_{v''_X}$ on the pressure of gas in N_2, $CO_2 - N_2$, and $CO_2 - N_2 - He$: 1) N_2; 2) $CO_2 - N_2 -$ He (1:3:8); 3) $CO_2 - N_2$ (1:2); 4) $CO_2 - N_2$ (2:1); 5) $CO_2 - N_2$ (9:1). The discharge took place in a tube of 20 mm diameter without cooling; $i_d =$ 20 mA.

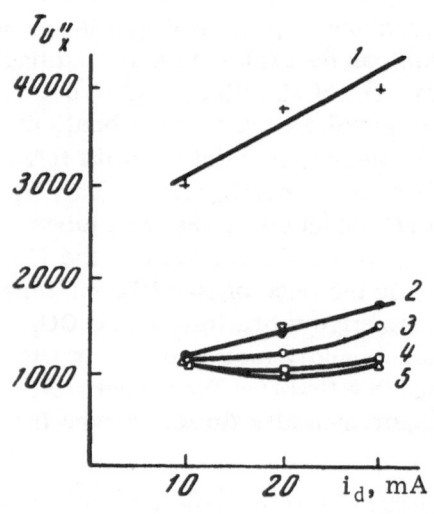

Fig. 37. Dependences of the vibrational temperatures $T_{v''_X}$ on the discharge current in various mixtures: 1) N_2 (p = 1.37 Torr); 2) $CO_2 - N_2 - He$ (1:3:8), p = 4.2 Torr; 3) $CO_2 - N_2$ (9:1); 4) $CO_2 - N_2$ (1:2); 5) $CO_2 - N_2$ (2:1), p = 2.1 Torr. The discharges took place in a tube of 20 mm diameter.

We shall not give a detailed interpretation of the dependences of the vibrational temperatures on the gas pressure and discharge current (Figs. 36 and 37). Changes in the gas temperature alter considerably the rates of relaxation processes. Clearly, an increase in the gas pressure and in the discharge current results, on the one hand, in an increase of the rate of excitation of the vibrational energy per unit volume and, on the other, in an increase of the gas temperature, i.e., an enhancement of the rate of deactivation of the vibrations. A simultaneous allowance for these (and also several other) factors is possible only in quantitative calculations whose results depend strongly on the gas temperature (see Chap. II). In our case the gas temperatures were not known because parallel measurements were not carried out.

When the present work was completed, Bleekrode [119] reported measurements of the vibrational temperature of nitrogen in a $CO_2 - N_2 - He$ mixture by a method similar to that described above. His results were in satisfactory agreement with ours.

CONCLUSIONS

The investigation reported in the present paper had the following results.

1. An optical method was developed for investigating the dissociation of CO_2 in the plasma of a CO_2 gas-discharge laser and this method was based on the absorption of infrared radiation.

2. This method was used in a study of the dissociation of carbon dioxide in discharge plasmas of continuous-flow and sealed CO_2 lasers as a function of the discharge current, pressure and composition of the gas mixture, rate of flow of the gas, diameter of the discharge tube, etc. Parallel measurements were made of the degree of dissociation and laser output power as a function of several parameters. The dependences of the dissociation rate constants on the discharge conditions were determined.

3. A system of kinetic equations, describing the processes of excitation and deactivation of vibrations in the multicomponent plasma of a CO_2 discharge laser, was derived and solved allowing for the dissociation processes. The population inversion densities were calculated for vibrational levels.

4. A method was developed for experimental determination of the vibrational temperatures in the ground electronic states of the N_2 and CO molecules, and of the antisymmetric vibrations of CO_2.

5. Measurements were made of the vibrational temperatures in $N_2 - CO_2$ and $N_2 - CO_2 - He$ gas mixtures.

A systematic experimental study of the dissociation of carbon dioxide in discharges and measurements of the output powers of CO_2 lasers combined with calculations and the results of measurements of the vibrational temperatures yielded the following conclusions.

The degree of dissociation of carbon dioxide in continuous-flow CO_2 lasers is considerable even in the case of systems with a relatively fast flow (this degree may be 60% or more).

Population inversion depends strongly on the degree of dissociation of CO_2. The CO molecules play an important role in the population of the upper laser level.

The electrode material and its treatment play an important role in the dissociation processes in sealed laser systems. The time dependence of the dissociation of CO_2 can be divided into several stages representing the establishment of a chemical equilibrium in the plasma and the interaction between plasma and the surrounding surfaces. The processes which limit the service life of sealed CO_2 lasers are identified. The addition of small amounts of hydrogen or water vapor improves the operation of sealed CO_2 lasers not only because of the deactivation of the lower laser level but also because of the catalytic activity of these additives in the dissociation-oxidation reactions.

The calculated vibrational inversion density is in agreement with the experimental results deduced from measurements of the gain (the agreement is to within a factor of 2).

In the case of continuous-flow laser systems the density of population inversion is distributed strongly inhomogeneously along the discharge tube because of the inhomogeneous chemical composition of the plasma.

The vibrational temperatures of the N_2, N_2-CO_2, and N_2-CO_2-He gases deduced from the distributions of the intensities in the nitrogen luminescence bands are in good agreement with the calculated values. This confirms the validity of the proposed interpretation of the main physical processes responsible for the population inversion in a CO_2 laser.

The author is deeply grateful to Professor N. N. Sobolev and Senior Scientist É. N. Lotkova for suggesting the subject, directing the work, and continuous advice.

The author is also grateful to E. S. Gasilevich, V. A. Ivanov, N. G. Yaroslavskii, and M. Z. Novgorodov for their valuable help.

APPENDIX I. MEASUREMENT OF GAS FLOW RATE

Figure 38 shows part of a gas supply system together with a discharge tube. The investigated mixture of gases was supplied from a cylinder 1 through a calibrated container 5 (of volume $V_5 = 2500$ cm^3) to a vacuum system 7 and a discharge tube 8. The positions of the valves 2-4 were such that the pressures p_5 in the container 5 ($p_5 \sim 2$ atm) and in the discharge tube 8 remained constant during measurements. This was possible because of the large reserve of gas in the cylinder 1 ($p_1 \sim 100$ atm, $V_1 \sim 3000$ cm^3).

The pressures in the vacuum system and in the discharge tube (p_8 is of the order of several torr) were measured with an oil manometer and the pressure in the container 5 was measured with a standard manometer 6.

The gas flow was determined by closing the valve 2 and measuring the rate of fall of the pressure $\Delta p_5/\Delta t$ in the container 5 subject to the condition $\Delta p_5 \ll p_5$. Since the same number of molecules in the gas mixture passed through each cross section in the system per unit time, we found that

$$p_5 V_5 = N_5 k T_5 \text{ (temp. outside discharge } T_5 = 300° \text{ K),}$$

$$\frac{dN_5}{dt} = \frac{V_5}{k T_5} \frac{dp_5}{dt} \text{ (N_5 is the number of particles in·container 5).}$$

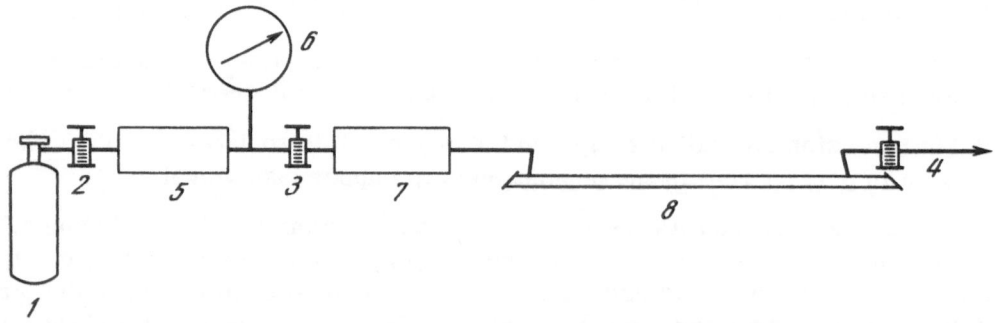

Fig. 38. System for measuring gas flow rate.

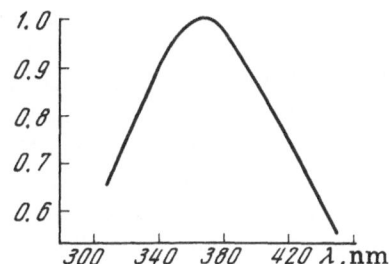

Fig. 39. Sensitivity curve of an FÉU-18-A photomultiplier.

Similarly,

$$\frac{dN_8}{dt} = \frac{p_8}{kT_8}\frac{dV_8}{dt},$$

where V_8, N_8, and T_8 are the volume, number of particles, and temperature in the discharge
tube. Since $dN_5/dt = dN_8/dt$, it follows that $dV_8/dt = (V_5/p_8)(T_8/T_5)(dp_5/pt)$. The linear rate of
flow (velocity) is $v = (1/S)(dV_8/dt)$, where S is the cross section of the gas-discharge tube.
The values of the discharge temperature T_8 can be taken from [43].

APPENDIX II. SPECTRAL SENSITIVITY OF THE PHOTOMULTIPLIER

The spectral sensitivity of the photomultiplier used in our measurements was determined
employing a standard ribbon lamp (SI-8-200), which was calibrated in terms of brightness
temperature at the effective wavelength $\lambda = 0.47\,\mu$. We measured the current in the lamp circuit and used a calibration graph to determine the brightness temperature T_B at $\lambda = 0.47\,\mu$.
The brightness temperature was then used to find the true temperature T of the tungsten ribbon
in the lamp and also the values of $T_B(\lambda)$ from the formula

$$\frac{1}{T_B} - \frac{1}{T} = \frac{\lambda}{c_2}\ln[\tau\varepsilon(\lambda,\,T)].$$

Here, τ is the transmission coefficient of the Uviol lamp window, $\varepsilon(\lambda,\,T)$ is the emissivity of
tungsten, and $c_2 = 1.438$ cm · °K. The values of $\varepsilon(\lambda,\,T)$ were taken from the de Vos paper [120].
Then, the true distribution of the intensity in the lamp spectrum was deduced from the Wien
law

$$b(\lambda,\,T) = \frac{c_1}{\lambda_5}\exp(-c_2/\lambda T_B),$$

where $b(\lambda,\,T)$ is the spectral energy brightness and $c_1 = 3.75 \cdot 10^{-5}$ erg · sec^{-1} · cm^2.

The sensitivity curve of the photomultiplier (Fig. 39) was plotted as a dependence of the
ratio of the ordinates of the true distribution of the intensities to the ordinates of the distribution obtained experimentally using this particular photomultiplier.

LITERATURE CITED

1.   C. K. N. Patel, W. L. Faust, and R. A. McFarlane, Bull. Am. Phys. Soc., 9:500 (1964).
2.   F. Legay and N. Legay-Sommaire, C. R. Acad. Sci. (Paris), 257:2644 (1963); 259:99
      (1964); N. Legay-Sommaire, L. Henry, and F. Legay, C. R. Acad. Sci. (Paris), 260:3339
      (1965); P. Barchewitz, L. Dorbec, R. Farreng, A. Truffert, and P. Vautier, C. R. Acad.
      Sci. (Paris), 260:3581 (1965); P. Barchewitz, L. Dorbec, A. Truffert, and P. Vautier,
      C. R. Acad. Sci. (Paris), 260:5491 (1965).

3.   T. G. Roberts, G. J. Hutcheson, J. J. Ehrlich, W. L. Hales, and T. A. Barr Jr, IEEE J. Quantum Electron., QE-3:605 (1967).

4.   W. B. Tiffany, R. Targ, and J. D. Foster, Appl. Phys. Lett., 15:91 (1969).

5.   N. N. Sobolev and V. V. Sokovikov, Usp. Fiz. Nauk, 91:425 (1967).

6.   V. P. Tychinskii, Usp. Fiz. Nauk, 91:389 (1967).

7.   C. K. N. Patel, Appl. Phys. Lett., 7:15 (1965).

8.   C. K. N. Patel, Phys. Rev. Lett., 13:617 (1964).

9.   N. N. Sobolev and V. V. Sokovikov, Proc. Fifth Intern. Conf. on Physics of Electronic and Atomic Collisions, Leningrad, 1967, p. 279; ZhETF Pis'ma Red., 4:303 (1966).

10. G. J. Schulz, Phys. Rev., 116:1141 (1959); 125:229 (1962); 135:A988 (1964).

11. N. N. Sobolev and V. V. Sokovikov, ZhETF Pis'ma Red., 5:122 (1967).

12. B. F. Gordiets, N. N. Sobolev, V. V. Sokovikov, and L. A. Shelepin, Relaxation Processes and Population Inversion in CO_2 Laser (Preprint) [in Russian], Lebedev Physics Institute, Academy of Sciences of the USSR, Moscow (1967); Phys. Lett. A, 25:173 (1967); B. F. Gordiets, N. N. Sobolev, and L. A. Shelepin, Zh. Eksp. Teor. Fiz., 53:1822 (1967); B. F. Gordiets, Thesis for Candidate's Degree [in Russian], Lebedev Physics Institute, Academy of Sciences of the USSR, Moscow (1968).

13. J. A. Macken, E. Matovich, and R. A. Brandewie, Bull. Am. Phys. Soc., 12:669 (1967).

14. É. N. Lotkova, V. I. Makarov, L. S. Polak, and N. N. Sobolev, Khim. Vys. Energ., 2:278 (1968).

15. A. B. Shekhter, Chemical Reactions in Electrical Discharge [in Russian], ONTI, Moscow-Leningrad (1935).

16. I. V. Nikitin and E. N. Eremin, Zh. Fiz. Khim., 36:616 (1962).

17. V. P. Glushko (ed.), Thermodynamic Properties of Individual Substances [in Russian], Izd. AN SSSR, Moscow (1962).

18. A. N. Mal'tsev, E. N. Eremin, and V. L. Ivanter, Zh. Fiz. Khim., 41:1190 (1967).

19. V. A. Sirotkina, Thesis for Candidate's Degree [in Russian], Moscow State University (1969).

20. I. A. Semiokhin, V. K. Korovkin, G. M. Panchenkov, and Kh. S. Bakhichevanskii, Zh. Fiz. Khim., 38:2072 (1964).

21. G. N. Meshkova and È. N. Eremin, Vestn. Mosk. Univ., Ser. Khim., No. 3, 8 (1966).

22. O. P. Bochkova and E. F. Shreider, Spectral Analysis of Chemical Mixtures [in Russian], Fizmatgiz, Moscow (1963).

23. P. K. Cheo and H. G. Cooper, IEEE J. Quantum Electron., QE-3:79 (1967).

24. É. N. Lotkova and N. Shukurov, Zh. Prikl. Spektrosk., 9:899 (1968).

25. R. J. Carbone, IEEE J. Quantum Electron., QE-3:373 (1967).

26. R. J. Carbone, IEEE J. Quantum Electron., QE-4:102 (1968).

27. W. J. Witteman and H. W. Werner, Phys. Lett. A, 26:454 (1968).

28. E. S. Gasilevich, V. A. Ivanov, É. N. Lotkova, V. N. Ochkin, N. N. Sobolev, and N. G. Yaroslavskii, Zh. Tekh. Fiz., 39:126 (1969).

29. J. E. Morgan and H. I. Schiff, J. Chem. Phys., 38:1495 (1963).

30. I. M. Campbell and B. A. Thrush, Proc. R. Soc. A, 296:201 (1967).

31. R. R. Reeves, G. Mannella, and P. Harteck, J. Chem. Phys., 32:632 (1960).

32. V. N. Kondrat'ev and I. I. Ptichkin, Kinet. Katal., 2:492 (1961).

33. M. A. A. Clyne and B. A. Thrush, Proc. R. Soc. A, 269:404 (1962).

34. M. A. A. Clyne and B. A. Thrush, Proc. Ninth Symposium on Combustion, Ithaca, N. Y., 1962, Academic Press, New York (1963), p. 177.

35. I. M. Campbell and B. A. Thrush, Trans. Faraday Soc., 64:1275 (1968).

36. P. Harteck and S. Dondes, J. Chem. Phys., 23:902 (1955).

37. F. H. R. Almer, M. Koedam, and W. M. ter Kuile, Z. Angew. Phys., 25:166 (1968).

38. A. L. S. Smith, Molecular Composition Changes in a Flowing $CO_2-N_2-He-H_2O$ Laser (Preprint), University of St. Andrews, Scotland; Phys. Lett. A, 27:432 (1968); J. Phys. D, 2:1129 (1969).

39. A. Sugiyama and H. Inaba, Phys. Lett. A, 28:120 (1968).

40. N. Djeu, T. Kan, and G. J. Wolga, IEEE J. Quantum Electron., QE-4:256 (1968).

41. F. M. Taylor, A. Lombardo, and W. C. Eppers, Appl. Phys. Lett., 11:180 (1967).

42. A. G. Mishchenko, L. M. Pavlova, V. P. Tychinskii, and T. A. Fedina, Elektron. Tekh. Ser. 3, No. 2, 47 (1968).

43. A. G. Sviridov, N. N. Sobolev, and G. G. Tselikov, ZhETF Pis'ma Red., 6:542 (1967).

44. A. H. von Engel, Ionized Gases, Clarendon Press, Oxford (1955).

45. A. G. Gaydon and I. R. Hurle, Shock Tube in High-Temperature Chemical Physics, Chapman and Hall, London (1963).

46. L. M. Brusilovskaya, M. Z. Novgorodov, A. G. Sviridov, and N. N. Sobolev, Preprint No. 32 [in Russian], Lebedev Physics Institute, Academy of Sciences of the USSR, Moscow (1969); M. Z. Novgorodov, A. G. Sviridov, and N. N. Sobolev, ZhETF Pis'ma Red., 8:341 (1968).

47. H. E. Evans and P. P. Jennings, Trans. Faraday Soc., 61:2153 (1965).

48. G. Glockler and S. C. Lind, The Electrochemistry of Gases and Other Dielectrics, Wiley, New York (1939).

49. P. K. Cheo and H. G. Cooper, IEEE J. Quantum Electron., QE-3:79 (1967).

50. G. Moeller and J. D. Rigden, Appl. Phys. Lett., 7:274 (1965); 8:69 (1966).

51. W. Espe, Materials of High-Vacuum Technology, Vol. 1, Metals and Metalloids, Pergamon Press, Oxford (1966).

52. K. Kawasaki, T. Sugita, and S. Ebisawa, J. Chem. Phys., 44:2313 (1966).

53. T. F. Deutsch and F. A. Horrigan, IEEE J. Quantum Electron., QE-4:972 (1968).

54. W. J. Witteman, IEEE J. Quantum Electron., QE-4:786 (1968); Phys. Lett. A, 26:454 (1968).

55. R. A. Paananen, Proc. IEEE, 55:2035 (1967); P. Bletzinger and A. Garscadden, Appl. Phys. Lett., 12:289 (1968).

56. P. O. Clark and J. Y. Wada, IEEE J. Quantum Electron., QE-4:263 (1968).

57. W. J. Witteman, Phys. Lett., 18:125 (1965); Philips Res. Rep., 21:73 (1966) W. J. Witteman and G. van der Goot, J. Appl. Phys., 37:2919 (1966); W. J. Witteman, J. Chem. Phys., 37:655 (1962).

58. F. Fischer, H. Küster, and K. Peter, Brennstoff Chem., 11:300 (1930).

59. R. R. Reeves Jr., P. Harteck, B. A. Thompson, and R. W. Waldron, J. Phys. Chem., 70:1637 (1966).

60. D. Rosenberger, Phys. Lett., 21:520 (1966).

61. É. N. Lotkova, V. I. Makarov, and T. P. Pyataeva, Khim. Vys. Energ., 3:476 (1969).

62. N. Karube, E. Yamaka, and F. Nayao, J. Appl. Phys., 40:3883 (1969).

63. V. N. Kondrat'ev, Free Hydroxyl [in Russian], GONTI (1939).

64. S. S. Penner, Quantitative Molecular Spectroscopy and Gas Emissivities, Addison, Wesley, Reading, Mass. (1959).

65. E. T. Gerry and D. A. Leonard, Appl. Phys. Lett., 8:227 (1966).

66. I. K. Babaev, A. T. Glazunov, V. P. Tychinskii, and S. N. Tsys', Elektron. Tekh. Ser. 3, No. 3, 36 (1968).

67. H. Statz, C. L. Tang, and G. F. Koster, J. Appl. Phys., 37:4278 (1966).

68. K. F. Herzfeld and T. A. Litovitz, Absorption and Dispersion of Ultrasonic Waves, Academic Press, New York (1959).

69. E. V. Stupochenko, S. A. Losev, and A. I. Osipov, Relaxation Processes in Shock Waves [in Russian], Nauka, Moscow (1965).

70. V. N. Kondrat'ev, Kinetics of Chemical Reactions in Gases [in Russian], Izd. AN SSSR, Moscow (1958).

71. R. N. Schwartz, Z. I. Slawsky, and K. F. Herzfeld, J. Chem. Phys., 20:1591 (1952).

72. F. I. Tanczos, J. Chem. Phys., 25:439 (1956).

73. L. D. Landau and E. Teller, Phys. Z. Sowjetunion, 10:34 (1936).
74. H. S. W. Massey and E. H. S. Burhop, Electronic and Ionic Impact Phenomena, Oxford University Press (1952).
75. A. Eucken and E. Nümann, Z. Phys. Chem. B. (Leipz.), 36:163 (1937).
76. E. E. Nikitin, Opt. Spektrosk., 9:16 (1960).
77. K. L. Wray, J. Chem. Phys., 36:2597 (1962).
78. J. D. Swift, Br. J. Appl. Phys., 16:837 (1965).
79. C. K. Rhodes, M. J. Kelly, and A. Javan, J. Chem. Phys., 48:5730 (1968).
80. C. B. Moore, R. E. Wood, B. L. Hu, and J. T. Yardley, J. Chem. Phys., 46:4222 (1967); J. T. Yardley and C. B. Moore, J. Chem. Phys., 46:4491 (1967).
81. P. K. Cheo, IEEE J. Quantum Electron., QE-4:587 (1968); Appl. Phys. Lett., 11:38 (1967).
82. D. Meyerhofer, IEEE J. Quantum Electron., QE-4:762 (1968).
83. L. O. Hocker, M. A. Kovacs, C. K. Rhodes, G. W. Flynn, and A. Javan, Phys. Rev. Lett., 17:233 (1966).
84. W. A. Rosser, A. D. Wood, and E. T. Gerry, IEEE J. Quantum Electron., QE-4:336 (1968).
85. F. Legay, J. Chim. Phys., 64:9 (1967).
86. R. C. Millikan, J. Chem. Phys., 38:2855 (1963).
87. J. E. Morgan and H. I. Schiff, Can. J. Chem., 41:903 (1963).
88. J. G. Parker, J. Chem. Phys., 41:1600 (1964).
89. M. J. W. Boness and G. J. Schulz, Phys. Rev. Lett., 21:1031 (1968).
90. R. D. Hake Jr and A. V. Phelps, Phys. Rev., 158:70 (1967).
91. S. S. Vasil'ev and E. A. Sergeenkova, Zh. Fiz. Khim., 40:2373 (1966).
92. E. T. Antropov, I. A. Silin-Bekchurin, N. N. Sobolev, and V. V. Sokovikov, Preprint No. 43 [in Russian], Lebedev Physics Institute, Academy of Sciences of the USSR, Moscow (1968); E. T. Antropov, I. A. Silin-Bekchurin, and N. N. Sobolev, Phys. Lett. A, 26:359 (1968).
93. S. Mathur and S. C. Saxena, Appl. Sci. Res., 17:155 (1967).
94. J. O. Hirschfelder, C. F. Curtiss, and R. B. Bird, Molecular Theory of Gases and Liquids, Wiley, New York (1954).
95. P. O. Clark and M. R. Smith, Appl. Phys. Lett., 9:367 (1966).
96. A. Dalgarno, in: Atomic and Molecular Processes (ed. by D. R. Bates), Academic Press, New York (1962), p. 643.
97. T. F. Deutsch, IEEE J. Quantum Electron., QE-3:151 (1967).
98. I. K. Babaev, A. T. Glazunov, and S. N. Tsys', Zh. Prikl. Spektrosk., 9:610 (1968); 10: 583 (1969).
99. J. Y. Coester and P. Vautier, Infrared Phys., 7:173 (1967).
100. A. Garscadden and P. Beltzinger, Proc. Ninth Intern. Conf. on Phenomena in Ionized Gases, Bucharest, 1969, Contributed Papers, publ. Editura Academiei RSR, Bucharest (1969), p. 251.
101. R. A. Crane and A. L. Waksberg, Appl. Phys. Lett., 10:237 (1967).
102. M. Z. Novgorodov, V. N. Ochkin, and N. N. Sobolev, Preprint No. 172 [in Russian], Lebedev Physics Institute, Academy of Sciences of the USSR, Moscow (1969).
103. A. D. Sakharov, Izv. Akad. Nauk SSSR. Ser. Fiz., 12:372 (1948).
104. T. N. Popova, Izv. Vyssh. Uchebn. Zaved. Fiz., No. 2, 44 (1958).
105. K. A. Weis, J. C. Kershenstein, and W. J. Thaler, J. Quant. Spectrosc. Radiat. Transfer, 9:885 (1969).
106. O. N. Glagoleva and S. S. Vasil'ev, Zh. Fiz. Khim., 43:1348 (1969).
107. W. M. Vaidya and K. C. Nagpal, Proc. Phys. Soc. Lond., 81:682 (1963).
108. I. P. Zapesochnyi and V. V. Skubenich, Opt. Spektrosk., 21:140 (1966).
109. D. J. Burns, F. R. Simpson, and J. W. McConkey, J. Phys. B, 2:52 (1969).
110. M. Jeunehomme and A. B. F. Duncan, J. Chem. Phys., 41:1692 (1964).
111. A. N. Vargin, Thesis for Candidate's Degree [in Russian], Moscow Engineering-Physics Institute (1971).

112. N. P. Carleton and O. Oldenberg, J. Chem. Phys., 36:3460 (1962).

113. V. P. Sychev, Uch. Zap. Kishinev. Univ., 55:37 (1960).

114. V. N. Egorov, L. N. Tunitskii, and E. M. Cherkasov, Zh. Prikl. Spektrosk., 8:479 (1968).

115. É. N. Lotkova and V. N. Ochkin, Zh. Prikl. Spektrosk., 11:739 (1969).

116. A. N. Terenin and G. G. Neuimin, Izv. Akad. Nauk SSSR. Ser. Khim, No. 5, 247 (1942).

117. N. Ya. Dodonova, Thesis for Candidate's Degree [in Russian], (1953); Dokl. Akad. Nauk SSSR, 98:753 (1954).

118. R. N. Zare, E. O. Larsson, and R. A. Berg, Mol. Spectrosc., 15:117 (1965).

119. R. Bleekrode, IEEE J. Quantum Electron., QE-5:57 (1969).

120. J. C. de Vos, Physca (Utr.), 20:690 (1954).

EXPERIMENTAL INVESTIGATION OF ELECTRICAL AND OPTICAL PROPERTIES OF THE POSITIVE COLUMN OF GLOW DISCHARGES IN MOLECULAR GASES*

M. Z. Novgorodov

Experimental (probe, high-frequency, and optical) methods were used in investigations of the electron energy distribution function, total electron density, and band intensities in glow discharges in pure molecular gases and their mixtures with He and Xe, as used in CO_2 lasers. An analysis was made of the influence, on the probe characteristics and their derivatives, of a cylindrical probe acting as an electron sink. A method was developed for determining the unperturbed distribution function. It was found that the electron energy distributions in mixtures of molecular gases were not Maxwellian. The functions obtained were used to calculate the rates of excitation of the vibrational levels of the molecules participating in population inversion.

INTRODUCTION

The study reported below forms part of a major series of theoretical and experimental investigations of the properties of discharge plasmas in CO_2 lasers being carried out in the Laboratory of Low-Temperature Plasma Optics at the Lebedev Physics Institute in Moscow. The present paper reports experimental investigations of the electrical and optical properties of the positive column in glow discharges in molecular gases and their mixtures at moderate pressures.

The CO_2 laser is the most widely used among the gas lasers. The output power and efficiency of cw gas-discharge CO_2 lasers are superior to the corresponding characteristics of the other gas lasers. The medium in which a population inversion is established is the positive column of a glow discharge in a mixture of CO_2 with other gases. The understanding and optimization of the operation of CO_2 lasers requires a knowledge of the elementary processes occurring in the discharge plasma and those giving rise to a population inversion of the molecular levels in CO_2. By the time the present study was started in the author's laboratory [1], a physically justified hypothesis had been put forward on the mechanism of population inversion in CO_2 lasers. However, this hypothesis was based on the approximate knowledge of the

---

* Candidate's dissertation defended at the P. N. Lebedev Physics Institute, Academy of Sciences of the USSR, Moscow, on October 4, 1971. The work was carried out under the direction of N. N. Sobolev.

parameters of the discharge plasma in CO_2 lasers. For example, one of the most important plasma parameters, the electron density, was only known inaccurately ($N_e = 10^9$-10^{11} cm^{-3}). Information on the electron energy and its distribution was taken from papers describing investigations carried out under similar discharge conditions but in different gases. A check of this hypothesis and the development of a full physical theory of CO_2 lasers are impossible without reliable information on the electron density and energy distributions. Therefore, one of the most important tasks in the investigation reported below was to obtain information on the density and energy distribution of electrons in discharges under conditions as close as possible to those during the operation of CO_2 lasers.

Until recently, attention has been concentrated on electrical discharges in metal vapors, inert gases, and their mixtures because of the use of these materials in gas-filled lamps, mercury rectifiers, and gas-discharge devices. Relatively little work has been done on discharges in molecular plasmas. The greater number of degrees of freedom of molecules makes their energy spectra more complex than those of atoms. Consequently, the electron energy is expended in the excitation of these complex spectra and this is reflected in the parameters of the electron gas. Therefore, discharges in molecular gases should differ in certain specific aspects from those in inert gases and metal vapors. In view of the higher probability of inelastic electron—molecule collisions, we may expect a nonequilibrium energy distribution of electrons.

The energy distribution of electrons can be studied on the basis of the transport equation. However, this equation can be solved only if we know the probabilities of many elementary processes involving collisions between electrons and molecules. At present, these probabilities are either not known at all or are known with a limited precision. Moreover, in the case of some processes such as the dissociation of molecules in discharges, not even the basic mechanism is known. Clearly, in this state of experimental knowledge, it would be very desirable to determine the electron distributions.

CHAPTER I

REVIEW OF LITERATURE ON ELECTRICAL AND OPTICAL PROPERTIES OF PLASMAS IN MOLECULAR DISCHARGES AND BASIC INFORMATION ON CO_2 LASERS

§ 1. Basic Information on CO_2 Lasers

Patel et al. [2] were the first to report, in 1964, the stimulated emission of $\lambda = 10.6$ μ radiation of about 1 mW power due to vibrational transitions in the CO_2 molecule. Later investigations, which were largely empirical, resulted in improved output power and efficiency. This became possible mainly because of the addition of nitrogen, helium, water vapor, and xenon, and also because of the optimization of the resonator structure. A detailed review of the work done on CO_2 lasers up to 1967 can be found in [1, 3].

The active medium of CO_2 lasers is the positive column of glow discharges established at moderate pressures (p = 1-20 Torr) and low currents (i_d = 10-50 mA) in tubes of 10-100 mm diameter. The consequences of the decomposition of CO_2 are avoided by forcing the gas to flow continuously through the discharge tube. In systems of this kind no special sophistication is needed to achieve a $\lambda = 10.6$ μ output power of 40-60 W from a discharge 1 m long and the efficiency of such systems can be as high as 20-30%. More complex CO_2 laser systems using higher flow velocities and pressures, transverse discharges, etc. have been used recently in order to increase the specific output power.

The first attempt to explain the operation of a CO_2 laser was made by Patel himself [4] for a discharge in a binary CO_2-N_2 mixture. He assumed that the population of the upper laser level of the CO_2 molecule was due to collisions with vibrationally excited nitrogen, whose vibrational levels were known to be in good resonance with the antisymmetric vibrations of the CO_2 molecule. He proved this hypothesis in experiments in which N_2 was excited separately and then mixed with CO_2. However, Patel's attribution of the high population of the vibrational levels of the ground electronic states of N_2 to electron−ion and atom−atom recombination processes and also due to multistage cascade transitions from excited electronic states was not justified.

Researchers in the present author's laboratory suggested a different physically well-grounded hypothesis on the mechanism of the CO_2 laser action [1]. This hypothesis was based on the experimental data of Schulz [5] on the probabilities of the excitation of molecular vibra-tions of N_2 and CO by electron impact. According to Schulz's data, the excitation cross sec-tions of these molecules can be very large ($3 \cdot 10^{-16}$ cm^2 for N_2 and $8 \cdot 10^{-16}$ cm^2 for CO) when the electron energy is 2-2.5 eV. Sobolev and Sokovikov [1] considered also the work of Swift [6] (this work will be discussed in detail in § 3) on the distribution of electron energies in the positive column of a glow discharge in N_2. Under conditions close to those in a CO_2 laser (for similar values of the product pd, where p is the pressure and d is the tube diameter), Swift [6] obtained average electron energies which were 2-2.5 eV and the energy distribu-tion was not Maxwellian. Sobolev and Sokovikov [1] assumed that the electrical parameters of plasmas in mixtures of N_2 and CO_2 did not differ greatly from those in pure N_2 and they showed that the high population of the vibrational levels of the N_2 molecule and of the upper laser level 00^01 of the CO_2 molecule can be explained by assuming direct excitation by electron impact. An analysis of the experimental and theoretical data [1] shows that the deactivation of the lower laser level 10^00 of the CO_2 molecule may be due to collisions of CO_2 with atoms and molecules.

The validity of the hypothesis of Sobolev and Sokovikov is supported by the fact that it can explain stimulated emission from mixtures without nitrogen (such an explanation is not possi-ble on the basis of Patel's hypthesis), when nitrogen is replaced with CO formed as a result of dissociation of CO_2.

These physical considerations were used by Gordiets, Sokovikov, and Shelepin [7] to calculate the population inversion in a CO_2 laser as a function of several parameters such as the discharge current, composition and pressure in the working mixture, etc. Gordiets et al. [7] assumed that the electron energy distribution was Maxwellian and ignored the chemical processes in the CO_2 laser plasma. The results of their calculations described correctly though qualitatively the experimental dependences and thus confirmed the main physical ideas on the CO_2 laser action. However, the calculations of Gordiets et al. [7] differed by an order of magnitude from the experimental data.

An important forward step in the understanding CO_2 laser action was the experimental investigation of the plasma composition reported in [8] and then in [9]. It was shown there that the degree of dissociation of carbon dioxide into CO and O could be considerable. Allowing for the dissociation of CO_2 and using our first measurements of the electron energy distribu-tion [10] as well as the experimental values of the electron density in discharges [11], Ochkin [9] calculated the vibrational population inversion in a CO_2 laser and obtained a satisfactory agreement with the experimental results [12].

§ 2. Elements of the Kinetic Theory of Plasmas

Electrons in the plasma of a positive gas-discharge column acquire their energy mainly from the electric field. They lose their energy by elastic and inelastic collisions with mole-cules and ions. Elastic collisions and Coulomb interactions cause electrons to be scattered in all directions so that their motion becomes random. Depending on the relative effects of these factors, we can have various electron velocity distributions ranging from highly directional to completely random. Calculations of the resultant electron velocity distributions can only be made using the transport equation.

The Boltzmann transport equation for the electron velocity distribution function is

$$\frac{\partial f}{\partial t} + \mathbf{v}\,\text{grad}_\mathbf{r}\,f + \frac{e}{m}\left(\mathbf{E} + \frac{1}{c}[\mathbf{vH}]\right)\text{grad}_\mathbf{v}\,f + \sum S = 0. \tag{1.1}$$

Here, $f(\mathbf{r}, \mathbf{v}, t)$ is the distribution function where $f\,d\mathbf{v}d\mathbf{r}$ represents the average number of electrons in a volume $d\mathbf{v}d\mathbf{r}$, $\mathbf{v}$ is the electron velocity, and $\mathbf{r}$ is the corresponding radius vector; $\mathbf{H}$ is the magnetic field. The electron density N_e, average electron energy $\bar{\varepsilon}$, and electron current $\mathbf{j}$ at a point $\mathbf{r}$ and at a moment t can be expressed in terms of the distribution function as follows:

$$N_e = \int f(\mathbf{r}, \mathbf{v}, t)\,d\mathbf{v}, \tag{1.2}$$

$$\bar{\varepsilon} = \frac{1}{N_e} \int \frac{mv^2}{2} f(\mathbf{v}, \mathbf{r}, t)\,d\mathbf{v}, \tag{1.3}$$

$$\mathbf{j} = \int e\mathbf{v}f(\mathbf{v}, \mathbf{r}, t)\,d\mathbf{v}. \tag{1.4}$$

The second term in Eq. (1.1) represents the change in the distribution function due to spatial inhomogeneities. The third term describes the interaction of electrons with external electric and magnetic fields. Finally, the quantity ΣS is known as the collision integral, which describes the change in the distribution function due to collisions of electrons with one another and also with other plasma particles, i.e., in general

$$\Sigma S = S_m^{el} + S_m^{in} + S_i + S_e, \tag{1.5}$$

where the right-hand side consists of terms describing the elastic (S_m^{el}) and inelastic (S_m^{in}) collisions of electrons with molecules, collisions of electrons with ions (S_i), and collisions of electrons with one another (S_e).

The integrodifferential equation (1.1) is nonlinear because of the electron−electron collisions and cannot be solved in any general form. However, in certain important special cases we can make assumptions that can yield the distribution function [13]. We shall consider some of the cases which are important from the theoretical point of view and which will be needed in discussing our own results later.

The equation (1.1) is simplified using the basic features of the behavior of electrons in a plasma. Under steady-state conditions ($\partial/\partial t = 0$) we may assume that the random (thermal) velocity of electrons is much higher than the directional velocity. Consequently, the distribution function is assumed to depend largely on the absolute value (modulus) of the velocity and not on its direction. In the isotropic case ($\mathbf{H} = 0$) and on condition that the spatial gradient of the distribution function is parallel to the electric field $\mathbf{E}$, the distribution function can be represented in the form

$$f(\mathbf{v}, \mathbf{r}, t) = f_0(v, \mathbf{r}, t) + \frac{\mathbf{v}}{v}\mathbf{f}_1(v, \mathbf{r}, t), \tag{1.6}$$

i.e., we may assume that the function has a spherically symmetric part f_0 and a directional part $\mathbf{f}_1$. This representation is valid in cases of small value of δ_{eff}, which is the effective fraction of the energy lost by an electron in one collision with a molecule.

When this assumption is made about the form of the distribution function, we can replace Eq. (1.1) by a system of coupled equations for the functions f_0 and $\mathbf{f}_1$. In these equations the part of the collision integral due to collisions between electrons is of the order of $\nu_e f_0$, where

$$\nu_e(v) = 2\pi N_e \frac{e^2}{m^2 v^3} \ln\left(1 + \frac{D^2 m^2 v^4}{e^4}\right) \tag{1.7}$$

is the electron−electron collision frequency; e and m are the charge and mass of an electron; D is the Debye screening radius. The remaining terms, which describe collisions between electrons and heavy particles, are of the order of $\delta_{\text{eff}} \nu f_0 = \delta_{\text{eff}}(\nu_m + \nu_i)f_0$, where ν_m and ν_i are the frequencies of collisions of electrons with molecules and ions, respectively. Depending on the relationship between ν_e and $\nu\delta_{\text{eff}}$, the distribution function f_0 is either governed by electron−electron collisions or by collisions between electrons and heavy particles.

If $\nu_e \gg \nu\delta_{\text{eff}}$ (strongly ionized plasma), we find that to within terms of the order of $\nu\delta_{\text{eff}}/\nu_e$, the symmetric part of the distribution function f_0 is Maxwellian:

$$f_0 = N_e \left(\frac{m}{2\pi k T_e}\right)^{3/2} \exp\left\{-\frac{mv^2}{2kT_e}\right\}. \tag{1.8}$$

Physically, this means that electron−electron collisions should establish a Maxwellian distribution in a time of the order of $1/\nu_e$. If $\nu_e \gg \nu\delta_{\text{eff}}$, this process is much faster than the transfer of energy to heavy particles.

The electron temperature T_e in Eq. (1.8) can be found from the energy balance equation

$$\frac{dT_e}{dt} + \delta_{\text{eff}}(T_e)\nu_{\text{eff}}(T_e)(T_e - T) = \frac{2e\mathbf{E}}{3kN}\mathbf{j}(T_e), \tag{1.9}$$

where T is the gas temperature, and the effective frequency of collisions with heavy particles can be expressed in terms of the distribution function:

$$\nu_{\text{eff}}(T_e) = \frac{4\pi m}{3N_e kT_e} \int_0^\infty v^4 \nu(v) f_0(v)\, dv. \tag{1.10}$$

For an infinite homogeneous plasma the electron density N_e in the distribution (1.8) can be found from the ionization balance equation

$$\frac{dN_e}{dt} + (\nu_{\text{rec}} - \nu_{\text{ion}}) N_e = 0, \tag{1.11}$$

where ν_{rec} and ν_{ion} are the averaged (over the distribution) total frequencies of electron ionization and recombination in a plasma.

The directional part of the distribution function $\mathbf{f}_1$ is readily derived [13]:

$$\mathbf{f}_1 = -\mathbf{u}\frac{df_0}{\partial v}, \tag{1.12}$$

where $\mathbf{u}$ is the velocity of directional motion of electrons deduced from the equation

$$\frac{\partial \mathbf{u}}{\partial t} + \nu(v)\mathbf{u} = \frac{e\mathbf{E}}{m}. \tag{1.13}$$

In the other limiting case when the electron−electron collisions are of little importance so that $\nu_e \ll \nu\delta_{\text{eff}}$ (weakly ionized plasma), but the inelastic collisions can be ignored ($S_m^{\text{in}} = 0$), we readily obtain the following solution for the function f_0:

$$f_0 = c \exp\left\{-\int_0^v \frac{mv\,dv}{kT + \frac{2e^2E^2}{3m\delta_{el}v^2(v)}}\right\},$$ (1.14)

where $\delta_{el} = 2m/M$ is the proportion of the energy lost in elastic collisions with molecules of mass M.

In a weak electric field, defined by the condition

$$E \ll E_p = \sqrt{\frac{3}{2}kT\frac{m}{e^2}\delta_{el}v^2},$$ (1.15)

we obtain from Eq. (1.14) a Maxwellian distribution with an electron temperature equal to the gas temperature T.

In a strong electric field, corresponding to a reversal of the inequality sign in Eq. (1.15), the form of f_0 can differ considerably from the Maxwellian function because the collision frequency $\nu(v) = Nv\sigma(v)$, where N is the number of molecules per unit volume, and the dependence $\nu(v)$ are both governed by the dependences of the elastic-collision cross sections on the velocity. In the special case of $\sigma(v) = $ const, we find from Eq. (1.14) the Druyvesteyn distribution

$$f_0 = c \exp\left\{-\frac{3m^2\delta_{el}}{8e^2E^2l^2}v^4\right\}.$$ (1.16)

The average energy for this distribution is

$$\bar{\varepsilon} \cong 0.6\,\frac{eEl}{\sqrt{\delta_{el}}},$$ (1.17)

where $l = v/\nu(v) = 1/\sigma N$ is the mean free path of electrons. Only if $\sigma(v) \propto 1/v$ do we find that Eq. (1.14) yields a Maxwellian distribution with an electron temperature governed by the electric field:

$$T_e = \frac{2e^2E^2}{3km\delta_{el}v^2}.$$ (1.18)

Allowance for inelastic collisions is easiest when only a small proportion of the electron energy is lost in such collisions. Such cases are encountered in molecular plasmas when electrons lose little energy in the excitation of rotational $(\varepsilon_i \sim 10^{-2}\text{-}10^{-4}$ eV$)$ and vibrational $(\varepsilon_i \sim 10^{-1}\text{-}0.5$ eV$)$ levels. In this case we can no longer assume that $\delta = $ const $= 2m/M$ but we must allow for the dependence of this quantity on the electron velocity, which is governed by the probabilities of inelastic collisions between electrons and molecules. The function f_0 is then of the form

$$f_0 = c \exp\left\{-\frac{3m^2}{2e^2E^2l^2}\int_0^v v^3\delta(v)\,dv\right\}.$$ (1.19)

This approach is justified in the case of moderate average electron energies.

At higher energies the losses of highly energetic electrons become significant and such electrons can excite optical levels or ionize molecules. This is true of inert gases for which the minimum excitation energy is usually at least 10 eV. Then, the inelastic-collision integral can be described by the limit formula

$$S_m^{in} = \nu^{in}(v)f_0, \tag{1.20}$$

where $\nu^{in}(v)$ is the frequency of inelastic collisions expressed in terms of the corresponding cross section. The solution of the transport equation below the excitation threshold is described by a Maxwellian or Druyvesteyn function and beyond the threshold we can use functions which decrease much more rapidly [13]. Thus, the "tail" of the distribution function in the range $\varepsilon > \varepsilon_i$, where ε_i is the excitation energy, seems to be cut off by inelastic collisions.

In the intermediate case in respect of ionization, when the function f_0 is affected strongly by electron–electron collisions as well as by collisions with heavy particles ($\nu_e \sim \delta_{eff} \nu$), the solution of the transport equation can be obtained by the iteration method selecting a Maxwellian distribution function with an electron temperature of Eq. (1.9) as the zeroth approximation. The ratio of the frequencies of electron–electron and electron–neutral collisions, given by the parameter

$$p = \nu_e(v_0)/\delta_{eff}\nu_m(v_0), \tag{1.21}$$

where $v_0 = \sqrt{2kT_e/m}$ and $T_e = eEl/\sqrt{6\delta_{eff}}$, governs the function f_0 and we can then distinguish the cases of strongly ($p \gg 5$) and weakly ($p \ll 5$) ionized plasma.

The treatment in [13] was improved considerably by Golubovskii, Kagan, and Lyagushchenko [14], who did not restrict in any way the ratio of the electron–electron and electron–neutral interactions and allowed also for the dependence of the elastic collision frequency on the energy. The distribution functions obtained in [14] have been used in calculations of electron mobilities in inert gases.

Earlier, Kagan and Lyagushchenko [15] solved the transport equation allowing for inelastic collisions in inert gases. They demonstrated that in the inelastic range the distribution function decreases much more rapidly than in the prethreshold range. Matching the solutions at the threshold energy $\varepsilon = \varepsilon_i$, Golubovskii et al. [16] obtained a distribution function which can be used to study the phenomenon of contraction of the positive column in neon discharges at moderate pressures.

Under molecular plasma conditions considered in the present paper we cannot apply the simplifications described above because at moderate electron energies both the vibrational levels of molecules and their electronic states are excited and their threshold energies are considerably lower than those of inert gases. Moreover, the form of the distribution function is affected by the dissociation of molecules.

Under these conditions it is usual to determine the symmetric part of the distribution function by numerical solution of the transport equation. One of the methods employed is the technique of successive approximations put forward in [17]. In these calculations the electron energy distribution function is an intermediate result which is used later to calculate the transport coefficients and to compare them with those found experimentally elsewhere. A repetition of this procedure for different cross sections gives more or less reliable information on the collision probabilities.

The distribution functions calculated for molecular nitrogen are given in [18] for three values of the reduced field E/N. It is assumed in these calculations that the plasma is infinite and homogeneous and no allowance is made for electron–electron collisions or for collisions of the second kind between excited molecules and electrons. It is clear from Fig. 1 (taken from [18]) that the electron energy distribution function is not Maxwellian and the deviation

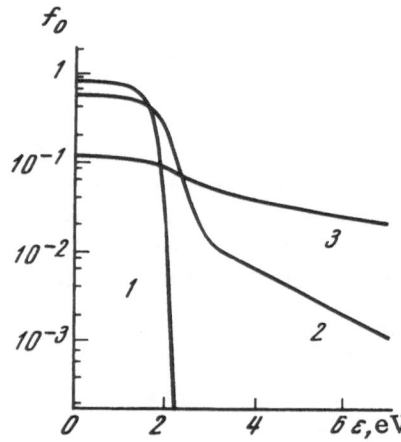

Fig. 1. Distributions of the electron energy $f_0(\varepsilon)$ in nitrogen [18] for $E/N = 1 \times 10^{-16}$ (1), 6×10^{-16} (2), and 30×10^{-16} V·cm² (3).

from the Maxwellian form is largely due to the excitation of the vibrational levels of the ground electronic state of the N_2 molecule. It should be pointed out that these calculations are based on the experimental values [5] of the excitation cross sections of the vibrational levels of N_2. The excitation process is characterized by a high probability. The maximum of the excitation cross section of the first eight vibrational levels of the N_2 molecule is $\sigma_{max} = 3 \cdot 10^{-16}$ cm² and it corresponds to the electron energy $\varepsilon = 2.3$ eV. Consequently, at this energy there is a sharp kink in the distribution function of the electron energies. Unfortunately, the distribution functions are not given in [19], reporting studies of the CO_2, CO, and O_2 molecules.

§ 3. Review of Experimental Measurements of Electrical Properties of Molecular Plasmas

When the investigation reported below was started there was only one published experimental paper of Swift [6], who measured the electron energy distribution functions under discharge conditions only remotely resembling those in a CO_2 laser. Swift determined the distribution function for the positive column of a glow discharge in nitrogen enclosed in a tube of 9 cm diameter and kept at pressures of p = 0.20-0.06 Torr (the discharge current was i_d = 100-400 mA). A differentiating circuit was used to obtain the first derivatives of the current produced by a spherical probe. The second derivatives, governing the distribution function, were obtained by graphical differentiation. The potential in free space (vacuum) was assumed to correspond to zero of the second derivative. It was found (Fig. 2) that the electron energy distribution was not Maxwellian and, moreover, it consisted of two groups of electrons with maxima at about 2 and 7 eV. The high-energy group of electrons represented an electron swarm which could form as a result of running losses. At the highest pressure (p = 0.2 Torr) employed in [6] the high-energy electron group was no longer observed and the distribution was neither Maxwellian nor Druyvesteyn. In fact, it was a narrow spectrum.

Thompson [20] investigated the energy distribution of electrons in the positive column in gas discharges in oxygen and nitrogen at low pressures p = 0.01-0.05 Torr and he used currents from 1 to 100 mA. Under these conditions there were strong standing losses and the electron energy distribution differed considerably from the Maxwellian form. The presence of several groups of electrons in the distribution was largely due to the loss characteristics, such as the dimensions of the regions where the losses occurred, magnitudes of the jump in the potential, etc. These characteristics themselves depended on the probabilities of inelastic collisions between electrons and gas molecules.

Garscadden and Bletzinger [21, 22] investigated the running losses in a positive column of discharges in N_2 and CO_2. They found that ionization waves excited in CO_2 and N_2 discharges

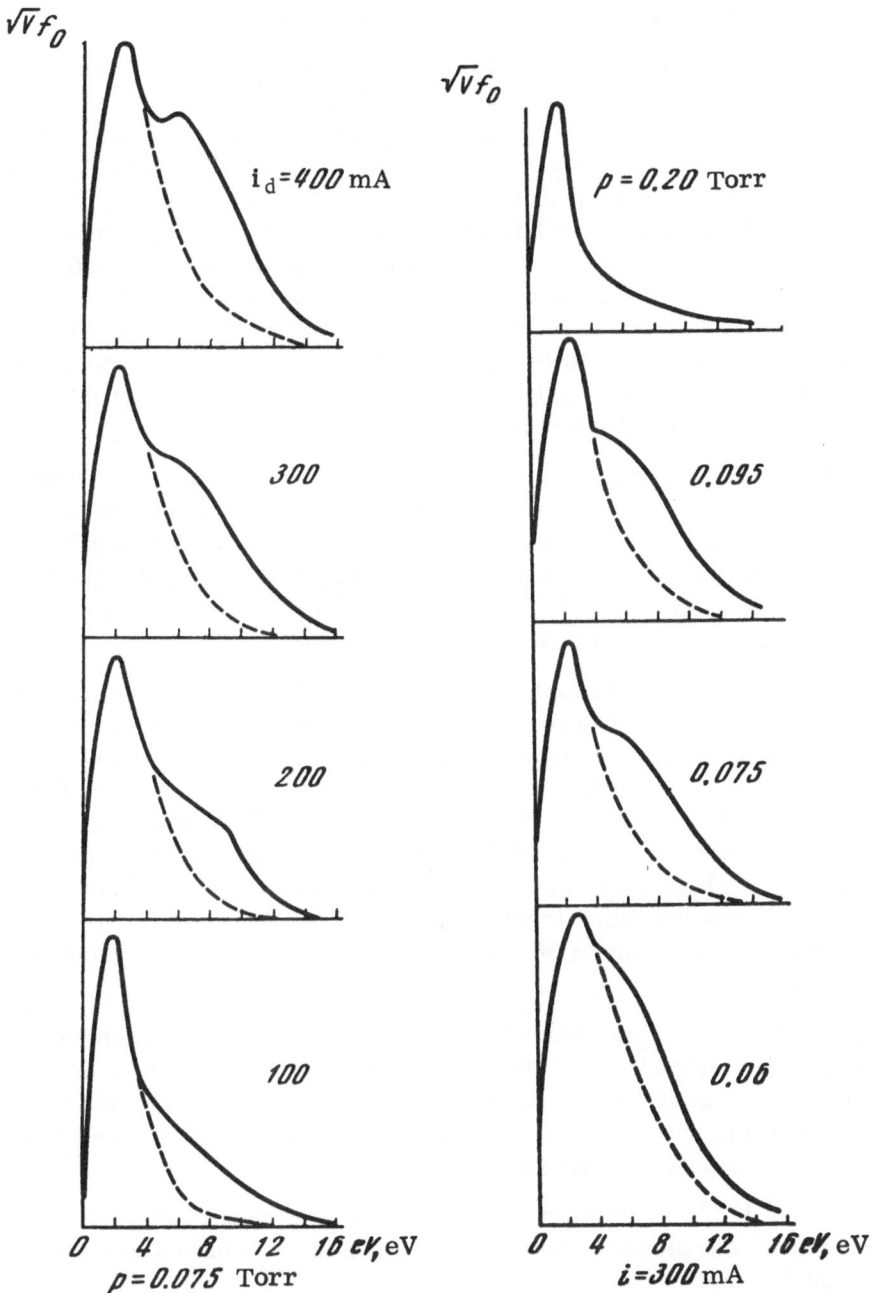

Fig. 2. Electron energy distributions in nitrogen discharges
$$f'(eV) = c\sqrt{V}f_0(eV) \quad [6].$$

were similar in respect of their kinetic and dispersion characteristics. At gas pressures exceeding 1 Torr we were unable to excite ionization waves and the discharge remained unperturbed. This was remarkable because our probe measurements were carried out at pressures exceeding 1 Torr. Unless special measures were taken, the distribution functions measured in the usual way under running loss conditions could be incorrect [23, 24].

Clark and Smith [25] were the first to investigate the parameters of a CO_2 laser plasma. They used a two-probe method in measurements of the electron temperature* and the distribution of the relative electron density; a microwave resonator was used in a determination of the absolute electron density as a function of the composition and pressure in the mixture. The measurements were carried out in a tube of 22 mm diameter using discharge currents of 10 and 15 mA. It was found that the addition of nitrogen to pure CO_2 at p = 1.5 Torr, so that the total pressure was about p = 4 Torr, reduced the electron temperature from 5 to 2.5 eV. The addition of He right up to 20 Torr to a mixture of 1.5 Torr CO_2 + 2 Torr N_2 maintained the electron temperature at 3 eV. The electron density decreased somewhat on addition of nitrogen to pure CO_2 and it amounted to about $5 \cdot 10^9$ cm^{-3}. Radial distributions of the relative values of the parameters indicated that the electron temperature was almost constant along the radius and the electron density had a Bessel-like distribution.

Carswell and Wood [26] investigated the properties of a CO_2 laser plasma and the influence of laser radiation on this plasma. The electron temperature was measured by a two-probe method and was found to be ~25,000-30,000°K in a tube of 5 cm diameter filled with a $He-N_2-CO_2$ (8 : 1 : 1) mixture at a pressure p = 5 Torr. The electron density determined using a microwave bridge operating at 17,000 MHz increased linearly from $3 \cdot 10^9$ to $20 \cdot 10^9$ cm^{-3} when the discharge current was increased from 40 to 200 mA.

Interesting results were obtained by Carswell and Wood on the dependences of the plasma properties on laser radiation. They found that the electron density and temperature as well as the emission of spontaneous radiation decreased in the presence of laser radiation, whereas the discharge resistance increased. These changes became greater when the discharge current was reduced. The discharge resistance was most sensitive to laser radiation and the changes in the resistance could reach 100%, so that a discharge could even stop when an optical resonator was opened. These experiments were remarkable because they demonstrated a strong interaction between the vibrationally excited molecules participating in the stimulated emission and the electron component of the plasma. Similar effects of laser radiation were considered in [27-29].

Properties of CO_2 laser plasmas were also investigated by Garscadden and Bletzinger [30-32]. In the first investigation [30], they used a one-probe method to determine the influence of the addition of Xe on the CO_2 laser parameters. They also determined the second derivatives of the probe current with respect to the voltage [31]. The influence of laser radiation on the probe characteristics was studied in [32].

Clark and Wada [33] investigated the influence of Xe on the properties of a sealed CO_2 laser. They measured the electron density by the resonator method and the electron temperature by the two-probe method.

Several microwave measurements were made of the radiation temperatures of discharges used in CO_2 lasers [34-37]. We shall analyze the results reported in [30-37] and compare them with our own results.

---

*We shall use here the term "electron temperature" following [25, 26]. A detailed discussion of this concept in the case of non-Maxwellian plasmas and criticism of [25, 26] are given in §5 of Chap. III.

CHAPTER II

EXPERIMENTAL METHODS AND CONDITIONS

§ 1.   Probe Method

General Basis of the Method

The probe method, suggested by Langmuir in 1923, is still one of the most widely used methods in the determination of plasma parameters.   There are several reviews and monographs [38–40] where the fundamentals of the probe method are given and references to earlier work are cited.   In this method an electric probe, i.e., a small electrode of spherical, cylindrical, or planar shape, is introduced into a discharge and subjected to a potential lower or higher than the plasma potential V_0 in the region under investigation.   If the probe is kept at a potential different from V_0, it produces in the surrounding space an electric field which accelerates charges of one sign and repels charges of the opposite sign.   In general, the total current i reaching a probe has the ion i_i and electron i_e components.

In electropositive gases (i.e., gases free of negative ions) the electron current to a probe is much greater than the ion current.   The theory of the electron current is much simpler and its results can be stated as follows.   The electron energy distribution function is independent of the distribution of the potential between the probe and an unperturbed plasma (in a repulsive potential) but it depends on the probe potential.   The density of the electron current reaching the probe is independent of the probe shape, provided it is convex, and it is given by

$$j_e = \frac{2\pi e N_e}{m^2} \int_{eV}^{\infty} (\varepsilon - eV) f_0(\varepsilon)\, d\varepsilon, \tag{2.1}$$

where e and m are the electron charge and mass, V is the probe potential, $\varepsilon = mv^2/2$ is the electron energy, and $f_0(\varepsilon)$ is the symmetric part of the distribution function.   Differentiating the above expression twice with respect to V, we obtain

$$\frac{d^2 j_e}{dV^2} = \frac{2\pi e^3}{m^2} N_e f_0(eV). \tag{2.2}$$

The formula (2.2) is the basis for the determination of the distribution function of the electron energies in an unperturbed plasma from the second derivative of the electron current with respect to the voltage.   If, instead f_0, we introduce the energy distribution function f' with the aid of the relationships

$$f'(\varepsilon)\, d\varepsilon = 4\pi v^2 f_0(\varepsilon)\, dv, \tag{2.3}$$

$$\varepsilon = \frac{mv^2}{2} = eV \tag{2.4}$$

[the quantity $f'(\varepsilon)d\varepsilon$ governs the number of electrons with energies between ε and $\varepsilon + d\varepsilon$], we find that $f'(\varepsilon)$ is

$$N_e f'(eV) = \frac{2\sqrt{2}}{e^2} \sqrt{\frac{m}{e}} \sqrt{V} \frac{d^2 j_e}{dV^2}. \tag{2.5}$$

If the electron distribution is Maxwellian (1.8), it follows from Eq. (2.1) that

$$j_e = \frac{N_e e \bar{v}_e}{4} \exp\left\{\frac{eV}{kT_e}\right\}, \tag{2.6}$$

where $\bar{v}_e = (8kT_e/\pi M)^{1/2}$ is the average velocity of electrons. This formula is used widely in plasma diagnostics. If we plot the dependence of $\ln j_e$ on V, we can use the slope of this dependence to determine the electron temperature

$$T_e = \frac{e}{k} \frac{dV}{d \ln j_e}. \qquad (2.7)$$

At the point where the probe potential is equal to the vacuum potential [V = 0 in Eq. (2.6)], the dependence of $\ln j_e$ on V has a kink. This kink can be used to find the electron density from the formula

$$N_e = \frac{4j_{e0}}{e\bar{v}_e} = \frac{4i_{e0}}{e\bar{v}_e S}, \qquad (2.8)$$

where i_{e0} is the electron current to the probe at the vacuum potential and S is the area of the probe surface.

Methods for Differentiation of Probe Characteristics

It is clear from Eq. (2.2) that we can obtain the electron energy distribution function from the second derivative of the probe current with respect to the voltage. Druyvesteyn [41] used Eq. (2.2) and graphical differentiation. His method suffers from serious shortcomings and it is very time-consuming. Simple differentiation methods [38] suffer from a considerable inaccuracy of the results obtained.

The second derivative of the probe current with respect to the voltage can be obtained using differentiating circuits. In this method the probe potential, measured relative to a selected electrode, is varied by applying the voltage from a sawtooth generator which ensures a linear time dependence of the voltage. The voltage drop across a small resistance in the probe circuit, proportional to the probe current, is passed to the input of the differentiating circuit. The output of this circuit is applied to the vertical plates of an oscillograph and the voltage from the sawtooth generator is applied to the horizontal plates. In this way the dependence of the first derivative of the probe current on the probe potential is displayed on the oscillograph screen. Sometimes the same method is used to find the second derivative [42].

An important improvement in the determination of the second derivative, compared with the earlier inaccurate graphical and semigraphical methods, is obtained by applying an alternating signal. In this method the probe circuit includes not only a source of a constant potential but also a source producing a small alternating signal v. Then, the probe potential is V + v and the current is

$$i = i(V + v). \qquad (2.9)$$

The probe characteristic without discontinuities can be represented by the Taylor series

$$i = i(V) + vi'(V) + \left(\frac{v^2}{2!}\right) i''(V) + \left(\frac{v^3}{3!}\right) i'''(V) + \ldots, \qquad (2.10)$$

where i is the probe current, V is the constant component of the probe voltage, i' = di/dV, i" = d^2i/dV^2, etc.

Depending on the form of the small alternating signal, we can distinguish several variants:

1) second harmonic method [43, 46]

$$v(t) = v_0 \sin \omega t; \qquad (2.11)$$

2) modulation of sinusoid method [45, 46]

$$v(t) = v_0 (1 + \cos \Omega t) \sin \omega t;$$ (2.12)

3) modulation of rectangular signal method [44]

$$v(t) = v_0 \left[1 + \frac{\pi}{4} \cos \omega t - \frac{\cos^3 \omega t}{3} + \cdots \right];$$ (2.13)

4) intermodulation method [46]

$$v(t) = v_{01} \sin \omega_1 t + v_{02} \sin \omega_2 t.$$ (2.14)

We used the second harmonic method. Substituting Eq. (2.11) into Eq. (2.10) and transforming the terms with the sines, we obtain

$$
\begin{aligned}
i = {} & i(V) + \frac{v_0^2}{4} i''(V) + \frac{v_0^4}{64} i^{IV}(V) + \frac{v_0^8}{2304} i^{VI}(V) + \cdots + \\
& + \left[v_0 i'(V) + \frac{v_0^3}{8} i'''(V) + \frac{v_0^5}{192} i^{V}(V) + \cdots \right] \sin \omega t - \\
& - \left[\frac{v_0^2}{4} i''(V) + \frac{v_0^4}{48} i^{IV}(V) + \frac{v_0^8}{1536} i^{VI}(V) + \cdots \right] \cos 2\omega t - \\
& - \left[\frac{v_0^3}{24} i'''(V) + \frac{v_0^5}{384} i^{V}(V) + \cdots \right] \sin 3\omega t + \\
& + \left[\frac{v_0^4}{192} i^{IV}(V) + \frac{v_0^8}{3840} i^{VI}(V) + \cdots \right] \cos 4\omega t + \cdots
\end{aligned}
$$ (2.15)

It is clear from the above formula that the harmonic frequency 2ω includes only the terms with even powers. If the amplitude of the alternating signal is small, the terms with higher derivatives can be ignored and then

$$i_{2\omega} = \frac{v_0^2}{4} i''(V) \cos 2\omega t.$$ (2.16)

It is clear from the above expression that we can find the second derivative of the probe current with respect to the voltage by measuring the amplitude of the second harmonic of the alternating signal.

An analysis of the errors resulting from dropping the amplitudes of the higher derivatives is given in [43]. It is found that the first three methods, described by Eqs. (2.11)-(2.13), are equivalent in respect of errors. The second harmonic method was used in [43] in the construction of an instrument for automatic recording of second derivatives.

An experimental comparison of the second harmonic, modulation of sinusoid, and intermodulation methods was carried out in [46] and it was found that the second harmonic and intermodulation methods had certain advantages and were more accurate than the modulation of sinusoid method.

Description of Apparatus and Analysis of Results

The electron energy distribution function was determined using apparatus which measured the second derivative of the probe current with respect to the voltage by the second harmonic method. A block diagram of the apparatus is shown in Fig. 3.

The probe characteristic and its second derivative were recorded relative to the anode of a discharge tube. The constant voltage drop in the discharge column between the probe and the anode was compensated by one or more series-connected stabilized voltage sources of the

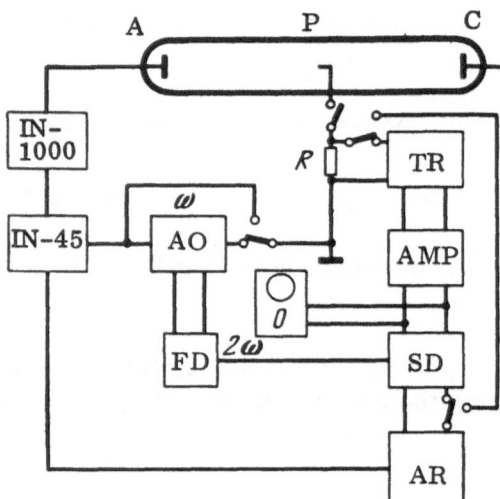

Fig. 3. Block diagram of the apparatus used in determination of the electron energy distribution function.

UIP-1, UIP-2, VS-11, or VS-12 type. A source of slowly varying voltage from 0 to 45 V (IN-45) was used for smooth variation of the voltage within the probe characteristic. This source was a potentiometer supplied from a stabilized voltage source capable of providing 45 V and 2 A. A rheostat taken from an ÉPP-09 instrument was used as the potentiometer; its resistance was 70 Ω. The potentiometer slider was rotated by a motor in such a way that one revolution of the slider was performed in 0.5-2 min, depending on the pulley diameter.

The alternating component of the probe voltage was supplied by a GZ-33 audiofrequency oscillator (AO in Fig. 3) with a 6-Ω output. The useful signal from the probe circuit was passed through a transformer TR to the input of a narrow-band amplifier AMP. The transformer TR was carefully screened and placed sufficiently far from the power supply to avoid picking up stray signals. The narrow-band amplifier was tuned to the second harmonic. The amplifier band width was variable and it was usually 1-1.5% of the working frequency. The amplified alternating signal of frequency 2ω was passed then to a synchronous detector SD. A reference signal was supplied to this detector from the oscillator through a frequency doubler FD, representing a double-half-period rectifier without a smoothing circuit. The shape and amplitude of the second harmonic signal was passed from the amplifier output to an oscillograph O. The amplifier and synchronous detector were standard instruments of the U2-6 type with SD-1 or KZ-2. The slowly varying second derivative signal was plotted by an automatic recorder AR, whose time constant for both coordinates was about 0.5 sec. The voltage from the IN-45 source was applied to the X input of the X-Y recorder. The reproducibility of the distribution function and its behavior in the tail region were determined by recording the second derivative several times using different amplification factors. We also included a schematic representation of the discharge tube in Fig. 3 (A is the anode, P is the probe, and C is the cathode).

The same apparatus was used in recording the one-probe characteristics. The audiofrequency oscillator was disconnected and the signal picked up from a resistance R, proportional to the probe current, was applied to the Y input of the X-Y recorder, bypassing the transformer, amplifier, and synchronous detector. The ionization frequency was determined by measuring the density of the ion current flowing to the discharge tube walls. This density was found by linear extrapolation of the ion current reaching a planar (flat) wall probe to the floating potential point where the total current to the probe was zero. A typical current-voltage characteristic, its second derivative, and ion current to a wall probe are plotted in Fig. 4.

The second derivative curves were analyzed on a computer. The values of the second derivative, obtained in steps of 0.2 or 0.25 V, were transferred to punched cards and a special-

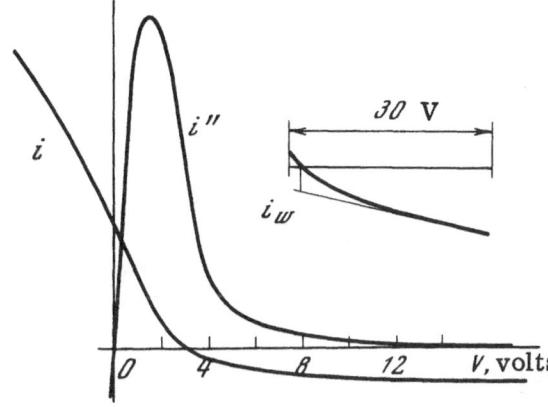

Fig. 4. Current-voltage characteristic of a probe (i), its second derivative (i"), and current flowing to the tube walls (i_w); N_2, p = 1.2 Torr, i_d = 40 mA.

ly prepared program carried out the following operations. The curve was normalized in accordance with the formula

$$\int_0^{\varepsilon_C} \sqrt{V}\, i''(V)\, dV = 1. \tag{2.17}$$

Then, the average electron energy was calculated from

$$\bar{\varepsilon} = \int_0^{\varepsilon_K} V^{3/2} i_n''(V)\, dV, \tag{2.18}$$

where i_n'' is the second derivative normalized in accordance with Eq. (2.17). Calculations were made of the average cross sections of inelastic processes, such as the excitation of the first eight vibrational levels of the N_2 and CO molecules, excitation of the vibrational level 00^01 of the CO_2 molecule, excitation of the electronic states $C^3\Pi_u$, $B^3\Pi_g$, and $A^3\Sigma_u^+$ of the N_2 molecule, and ionization of the ground states of the N_2, CO, and CO_2 molecules and of the Xe atom. These cross sections were calculated using the formula

$$\langle \sigma v \rangle = \sqrt{2/m} \int_{\varepsilon_n}^{\varepsilon_C} \sigma(V)\, V i_n''(V)\, dV, \tag{2.19}$$

where $\sigma(V)$ are the cross sections of inelastic collisions taken from [47-50]. The program made it possible to calculate normalized Maxwellian and Druyvesteyn distributions with the same average energy as the experimental distribution. The computer then plotted all the curves on a wide paper chart.

Possible errors of the differentiating circuit were checked by computing the probe characteristic using Eq. (2.1) and comparing it with the experimentally determined current–voltage characteristic.

Errors in Second Harmonic Method

It is clear from Eq. (2.15) that the second derivative of the probe current with respect to the voltage can be determined when the terms in the Taylor series with higher derivatives (third or higher) are small compared with $v_0^2 i''(V)/4$. This applies both to the second harmonic method and to the method used by Sloane and MacGregor [51] to measure the constant component of the probe current. Dropping terms with higher derivatives (beginning from the sixth) from

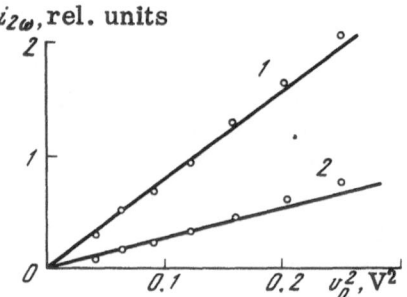

Fig. 5. Dependences of the amplitude of the second harmonic $i_{2\omega}$ on the square of the amplitude of the alternating signal v_0^2.

the expression for the second harmonic amplitude, we obtain

$$i = i_{2\omega} \cos 2\omega t \simeq \left[\frac{v_0^2}{4} i''(V) + \frac{v_0^4}{48} i^{IV}(V) \right] \cos 2\omega t, \tag{2.20}$$

where $i_{2\omega}$ is the second harmonic amplitude and v_0 is the amplitude of the alternating signal of frequency ω. The smallness of the term with the fourth derivative means that the second harmonic amplitude varies linearly with the square of the amplitude of the probe signal. We checked this prediction in several cases. Figure 5 shows the dependence of the second harmonic amplitude on v_0^2 for two values of the probe potential V_0. It is clear from this figure that the points nearly fit a straight line. Deviations from linearity are observed at alternating signal amplitudes exceeding 0.4 V. Therefore, in all the subsequent experiments we ensured that the alternating signal amplitude did not exceed 0.4 V.

Call [52] studied theoretically the errors in the second harmonic method by Fourier analysis and found that the distortion of the second derivative as a result of superposition of the alternating signal depends on the ratio $v_0/\bar{\varepsilon}$, where $\bar{\varepsilon}$ is the average electron energy. [This analysis was made by Call [52] for a current−voltage characteristic represented by the function $i(V) = \tanh(V/\bar{\varepsilon})$. This hyperbolic tangent expression described satisfactorily many experimentally determined characteristics.] For ratios $v_0/\bar{\varepsilon} < 0.2$ the distortions were slight and they occurred mainly in the region of rapid changes in the second derivative, i.e., in the region of the maximum and zero of i''. We shall show later that the average energies in our electron distributions are about 2–3 eV. Therefore, the selected amplitude of the alternating voltage component $v_0 = 0.4$ V can be regarded as satisfactory from the point of view of the requirements set out in [52].

Probe as an Electron Sink

A probe absorbing all electrons perturbs the surrounding plasma. This perturbation can easily be estimated. Clearly, a probe reduces locally the electron density. We can show that this reduction is governed by the ratio of the mean free path of electrons λ_e and the probe radius a. For example, the current density j_0 for the vacuum potential is [40]

$$j_0 = \frac{N_e e \bar{v}_e}{4(1 + 3\lambda_e/4a)} \tag{2.21}$$

for a spherical probe and

$$j_0 = \frac{N_e e \bar{v}_e}{4\left(1 + \frac{3}{4} \frac{\lambda_e}{a} \ln \frac{l}{a}\right)} \tag{2.22}$$

for a cylindrical probe of length l.

The loss of electrons in the vicinity of a probe [compare Eqs. (2.21) and (2.22) with Eq. (2.7)] should be restored by diffusion from the surrounding space. Since the rate of diffusion is proportional to the electron velocity, this process will occur at different rates for electrons of different energies. Thus, in obtaining the second derivative a probe records a distribution function different from that found in an unperturbed plasma.

Swift [53] was the first to study the distortion of the distribution function due to the sink effect at a probe. He obtained an expression for the current flowing to a spherical probe and for the second derivative of this current in the case of an arbitrary distribution function. The main parameter which governs the distortion is

$$\psi = \frac{3}{2} \frac{(a/\lambda_e)^2}{1 + a/\lambda_e}. \tag{2.23}$$

Swift calculated the distortions for the Maxwellian and Druyvesteyn distributions and he indicated the limits of the probe radius, pressure (governing the mean free path of electrons), and electron energy within which the distortions did not exceed 25%.

Kagan and Perel' [54] solved the same problem in a different way. The distribution function of electrons arriving on the surface of a spherical probe was found by solving the transport equation subject to the condition that the distribution function was Maxwellian at infinity.

In modern probe measurement practice it is usual to employ cylindrical rather than spherical probes. Therefore, it is desirable to obtain quantitative estimates of the distorting influence of the sink effect on the probe characteristics and their derivatives in the cylindrical case.

It was shown by Lukovnikov and the present author in [55] that the current reaching a cylindrical probe can be described by the expression

$$i = 2\pi a a l \int\limits_{eV}^{\infty} f(\varepsilon)(\varepsilon - eV)\, d\varepsilon, \tag{2.24}$$

where $\alpha = \frac{1}{4} N_e (2e/m)^{1/2}$; $\varepsilon = mv^2/2$ is the electron energy; V is the probe potential. This relationship is derived in [55] for a cylindrical probe by a method different from that used in [56] and it can be used conveniently to find the change in the number of particles on the collecting surface due to the loss of electrons to the probe:

$$di = 2\pi a a l f(\varepsilon)(\varepsilon - eV)\, d\varepsilon. \tag{2.25}$$

[The distribution function at the boundary of a layer $f(\varepsilon, r)$ is assumed to be isotropic and normalized to unity $\int\limits_0^\infty f(\varepsilon, r)\varepsilon^{1/2}d\varepsilon = 1$]. On the other hand, the replacement of electrons by a diffusion flux can be described by

$$di = 2\pi r l\,(1 + 2r/l)\, D\, \frac{d}{dr}\,[dN_e(r, \varepsilon)], \tag{2.26}$$

where $D = \frac{1}{3}\lambda_e \varepsilon^{1/2}(2e/m)^{1/2}$ is the diffusion coefficient; $dN_e(r, \varepsilon) = N_e \varepsilon^{1/2} f_l(\varepsilon, r)d\varepsilon$ is the number of electrons in an energy interval $d\varepsilon$ at a distance r from the probe axis. The factor $(1 + 2r/l)$ allows for the current through the ends of the layer and its introduction avoids divergences during the subsequent integration. Integrating Eq. (2.26) with respect to r between the limits b (representing the layer boundary) and ∞, we obtain

$$f(\varepsilon)\, d\varepsilon = f_0(\varepsilon)\, d\varepsilon - \frac{di}{2\pi \lambda_e D \varepsilon^{1/2}}\, \ln\,(1 + l/2b), \tag{2.27}$$

where $f_0(\varepsilon)$ is the electron distribution function far from the probe. It follows from the above equality that the distortion of the distribution function of the electrons reaching the probe surface is manifested primarily by a reduction in the number of the low-energy electrons. This is manifested by the minus sign in front of the term representing diffusion and by the inverse proportionality of this term to the electron energy. Distortions decrease when the electron energy or the mean free path λ_e are increased.

Substituting $f_0(\varepsilon)$ from Eq. (2.27) into Eq. (2.25) and integrating with respect to the energy, we obtain the following expression for the dependence of the density of the current flowing to a cylindrical probe on the potential V:

$$j = \frac{i}{2\pi a l} = a \int_{eV}^{\infty} \frac{f_0(\varepsilon)(\varepsilon - eV)\,d\varepsilon}{1 + \delta(1 - eV/\varepsilon)}. \tag{2.28}$$

Differentiating this expression twice with respect to V, we find that

$$j'' = af_0(\varepsilon)\big|_{\varepsilon = eV}[1 - \theta(f_0)], \tag{2.29}$$

where

$$\theta(f_0) = 2 \int_{eV}^{\infty} \frac{\delta f_0(\varepsilon)\,d\varepsilon}{\varepsilon f_0(eV)[1 + \delta(1 - eV/\varepsilon)]^3}, \tag{2.30}$$

$$\delta = \frac{3}{4}\frac{a}{\lambda_e}\frac{a}{b}\ln\left(1 + \frac{l}{2b}\right) \simeq \frac{3}{4}\frac{a}{\lambda_e}\ln\frac{l}{2a}. \tag{2.31}$$

The expressions for the current reaching a spherical probe [53, 54] are similar to Eq. (2.28) if the quantity δ is suitably defined. It should be noted that, in contrast to [54], the unperturbed distribution function f_0 in Eq. (2.28) need not be Maxwellian.

The effect under consideration distorts strongly the function f_0, particularly at low energies and for high values of δ. For example, in the case of the Maxwellian function $f_0 = c \exp\left\{-\frac{\varepsilon}{eV}\xi\right\}$ $\left(\delta(\varepsilon) = \text{const}, \xi = \frac{eV}{kT}\right)$ we can show that if $\delta \ll 1$, then

$$\theta \simeq 2\delta E_1(\xi) \quad \text{for} \quad \xi \ll 1, \quad \theta \simeq 2\delta/\xi \quad \text{for} \quad \xi \gg 1,$$

whereas if $\delta \gg 1$, then

$$\theta = 1 + \frac{2}{\delta^2}E_1(\xi/\delta) > 1 \quad \text{for} \quad \xi/\delta \ll 1, \quad \theta = (1 - 2/\delta) \simeq 1 \quad \text{for} \quad \xi/\delta \gg 1,$$

where $E_1(x) = e^{-x}\int_0^{\infty}\frac{e^{-z}}{x+z}\,dz$.

In an analysis of the experimental results it is essential to determine the function $f_0(eV)$ from the measured second derivative of the probe current density j''. In general, this problem can be solved by representing Eq. (2.29) as an integral equation for the function f_0:

$$f(eV) = f_0[1 - \theta(f_0)]. \tag{2.32}$$

Equation (2.32) can be solved by the method of successive approximations, using the experimental function $f_0^{(0)} = j''$ as the zeroth approximation:

$$f_0^{(i+1)} = f_0^{(i)}/[1 - \theta(f_0^{(i)})]. \tag{2.33}$$

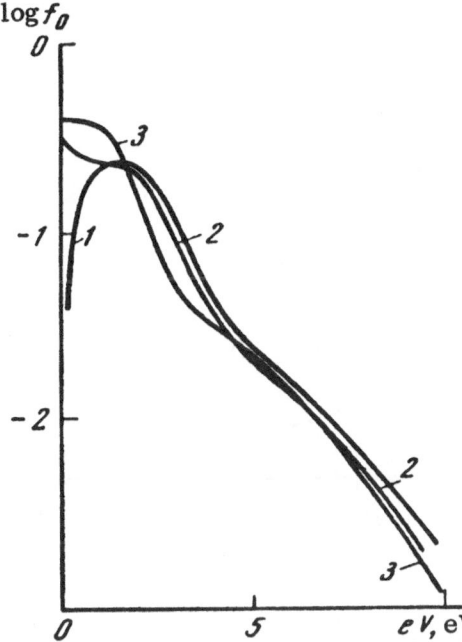

Fig. 6. Electron energy distributions: 1) $\log i''$, $\bar{\varepsilon} = 2.8$ eV, N_2, p = 1 Torr, E/N = 10^{-15} V·cm^2; 2) $\log i''/(1 - \theta)$, $\bar{\varepsilon} = 2.5$ eV, N_2, p = 1 Torr, E/N = 10^{-15} V·cm^2; 3) $\log f_0$, $\bar{\varepsilon} = 2.4$ eV, N_2, E/N = 10^{-15} V·cm^2 [58].

The smaller the value of δ, the faster the convergence of the solution for the bounded function $f(\varepsilon)$, where $\varepsilon \neq 0$ [57]. Figure 6 shows the experimentally determined second derivative of the probe current and the first approximation $j''/(1 - \theta)$. Introduction of a correction removes the fall of the distribution function at low electron energies. For comparison, Fig. 6 includes the calculated distribution function obtained in [58].

The expression for the electron current flowing to a probe at the vacuum potential, i.e., Eq. (2.7), is used to determine the electron density. We can see from Eq. (2.28) with V = 0 that as a result of the electron sink effect the measured value of N may be underestimated by a factor exceeding $(1 + \delta)$. The formula (2.29) makes it possible to express the ratio of the density N_{e0} (related to the unperturbed distribution function f_0) to the electron density N_e:

$$\frac{N_{e0}}{N_e} = 1 - 0.5 \left(\frac{1 + 2\delta}{1 + \delta} + \frac{0.5}{\delta^{1/2}(1 + \delta)^{3/2}} \ln \frac{(1 + \delta)^{1/2} - \delta^{1/2}}{(1 + \delta)^{1/2} + \delta^{1/2}} \right). \qquad (2.34)$$

Here, we find that $N_{e0} \approx (1 + 4\delta/3)N_e$ for $\delta \ll 1$ and $N_{e0} = 2\delta N_e$ for $\delta \gg 1$. The dependence of N_{e0}/N_e on δ is plotted in Fig. 7.

The average electron energy is overestimated as a result of the sink effect. Using Eq. (2.29), we can obtain the ratio of the average energy for the unperturbed distribution $\bar{\varepsilon}_{e0}$ to the average energy for the function found experimentally $\bar{\varepsilon}_e$ as a function of the parameter δ:

$$\frac{\bar{\varepsilon}_{e0}}{\bar{\varepsilon}} = \frac{N_e}{N_{e0}} \bigg/ \left(\frac{3}{2\delta} + \frac{3}{4\delta^{3/2}(\delta + 1)^{1/2}} \ln \frac{\sqrt{1 + \delta} - \sqrt{\delta}}{\sqrt{1 + \delta} + \sqrt{\delta}} \right). \qquad (2.35)$$

If $\delta \gg 1$, we find that $\bar{\varepsilon}_{e0}/\bar{\varepsilon} \approx 3$ but if $\delta \ll 1$ then $\bar{\varepsilon}_{e0} = \bar{\varepsilon}_e$. Figure 7 shows the dependence of $\bar{\varepsilon}_{e0}/\bar{\varepsilon}_e$ on δ.

In an analysis of the experimental results on a computer the electron sink effect is readily allowed for by means of Eqs. (2.28)–(2.31). Therefore, a procedure allowing for this effect was introduced into the computer program.

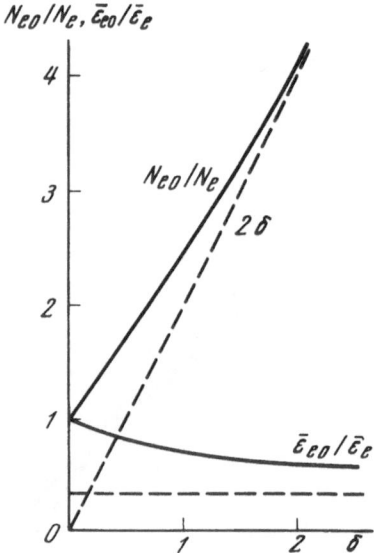

Fig. 7. Dependences of N_{e0}/N_e and $\bar{\varepsilon}_{0e}/\bar{\varepsilon}_e$ on the parameter δ. Here, N_{e0} and $\bar{\varepsilon}_{e0}$ are the density and average energy of electrons in an unperturbed plasma; N_e and $\bar{\varepsilon}_e$ are the density and average energy of electrons deduced from the measured second derivative. The dashed lines are the asymptotes.

Vacuum Potential in Probe Measurements

According to the classical theory of probes the plasma potential corresponds to the point of inflection of the curve current representing the voltage dependence of the electron current density because the first derivative should have a maximum and the second a discontinuity (Fig. 8a). However, the experimentally determined characteristics differ from those discussed above. In particular, an inflection of the experimental curve j_e is smoother and the curve $j_e'' = d^2 j_e/dV^2$ does not have a discontinuity. One of the typical experimental curves is shown in Fig. 8b.

The separation between zero and the maximum of the second derivative depends on the discharge conditions and probe dimensions, and it may reach considerable values. Under our conditions this separation is 1.5-1.5 V.

Until recently the experimentalists have not accepted the same and suitably justified view on the analysis of the second derivative curves. Some investigators have plotted the functions $f(\varepsilon)\varepsilon^{1/2}$ measuring the energy from the point where $d^2 j_e/dV^2 = 0$, whereas others measured the energy from the maximum of the second derivative. It should be pointed out that this indeterminacy of the selection of the point corresponding to the vacuum potential may give rise to variations in the form of the distribution function which governs the degree of equilibrium of the electron component, efficiencies of various processes in a plasma, frequencies of collisions with other components of the plasma, average electron energies, etc. The differences may be particularly large in the electron density deduced from the electron

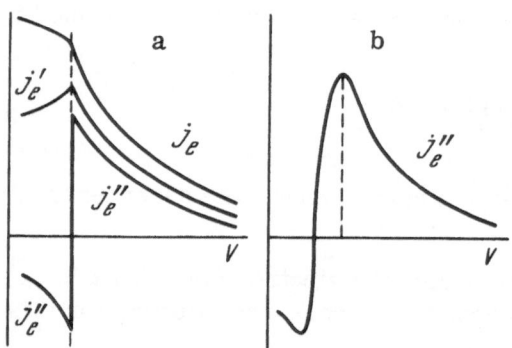

Fig. 8. Schematic representation of the probe characteristics and their derivatives for a cylindrical probe: a) ideal characteristic j_e, and its first j_e' and second j_e'' derivatives; b) typical experimental second derivative j_e''.

current reaching a probe for a given value of the vacuum potential. This is due to the fact that, in the range under consideration, the electron current depends strongly on the potential and a relatively small separation between zero and the maximum of i'' may give rise to values of the electron current differing severalfold.

Calculations employing the transport equation (1.1) at the point $\varepsilon = 0$ give finite values for the symmetric part of the distribution function (see, for example, Figs. 1 and 6). Therefore, the choice of the vacuum potential at zero of the second derivative (j'' is an analog of the symmetric part of the distribution function) is not justified from the point of view of theoretical distributions.

Spreading of the second derivative curve near the vacuum potential may be due to the reflection of electrons from a probe [38] and the presence of a high-frequency component in the probe potential [59]. The available information [60] on the reflection coefficients shows that they are small and practically independent of the electron energy. Call's calculations [52] demonstrate that the usual amplitudes of the alternating signal result in a spread of the second derivative curves which is much less than that observed. Therefore, it is clear that these two factors cannot explain the real form of the second derivative curves near zero and maximum values.

Vorob'eva et al. [59] suggested a method for selecting the vacuum potential at the maximum of i'' on the basis of the best agreement between the electron temperature deduced from $T_e = {}^2/_3 \bar{\varepsilon}$ and from the position of the maximum of $\varepsilon^{1/2} f_0(\varepsilon)$ ($\varepsilon_{max} = T_e/2$) for the Maxwellian distribution function $f_0(\varepsilon) = \exp(-\varepsilon/T_e)$. Since the second derivative for the Maxwellian function is the same as the function itself, we should have $-\exp(-\varepsilon/T_e)$, which is as expected because the position of the maximum of $\varepsilon^{1/2} f_0(\varepsilon)$ is independent of the point from which energy is measured but is simply governed by the electron temperature T_e. Clearly, the selection of the energy origin at the maximum of i'' gives better agreement than the selection at zero of i''. Therefore, the method for finding the vacuum potential suggested in [59] cannot be regarded as justified.

However, there are also arguments in support of the selection of the vacuum potential at zero of the second derivative. It follows from the author's study [55] that the nature of the probe characteristic and, consequently, the positions of zero and maximum of the second derivative depend not only on the effects listed above but also on the pressure, probe geometry, and electron energy. It is clear from Eq. (2.29) that the second derivative of the probe current vanishes at $\theta(f_0) = 1$. This occurs when the probe is still negative relative to the plasma. Thus, the true vacuum potential, at least in the presence of the electron sink effect, lies in the range of negative values of the second derivative and not near the maximum. It follows from Eqs. (2.29)–(2.31) that in the case of an unperturbed Maxwellian distribution function and $\delta \ll 1$ the vacuum potential is shifted relative to zero of i'' by $\Delta V = 0.56\, T_e \exp(-\delta/2)$. If $\delta \gtrsim 1$, this shift is much greater.

Recent papers have provided other indirect arguments in support of the selection of the vacuum potential at zero of the second derivative. For example, a sudden drop in the distribution function at the first excitation potential and a comparison with tabulated values in [61] indicate that the vacuum potential should be selected as suggested here. Measurements of the noise amplitude [62] have shown that the noise maximum occurs at the zero of the second derivative, which again supports [63] the proposed selection of the vacuum potential.

Our comparison of the electron density [64] determined from the current to the probe given by Eq. (2.7), obtained for different values of the vacuum potential subject to the sink effect with the densities deduced from microwave measurements, demonstrates a good agreement with the values found at zero of the second derivative. Similar conclusions follow also from the results reported in [65].

In the present study the position of the vacuum potential was found by analysis of the experimental data in accordance with Eqs. (2.32) and (2.33). In most cases its position was found to be practically identical with zero of the second derivative (the range of ΔV was 0.1-0.3 V).

§ 2. High-Frequency Resonator Method

Basis of the Method

It is not possible to provide a full description of the physical processes resulting in population inversion in CO_2 lasers without knowledge of the absolute electron density. Stimulated emission from these lasers is usually obtained as a result of discharges in multicomponent mixtures. Naturally, the ion composition of the plasma cannot be determined without special investigations. Therefore, we cannot use the ion component of the current−voltage probe characteristics to find the electron density because the density of the ion current reaching a probe depends on the ions mass [40]. It is also clear from the preceding section that the use of the electron part of a characteristic obtained at moderate pressures is also a doubtful procedure. Therefore, it is desirable to obtain information on the absolute electron density by an independent method. We used the high-frequency resonator method for this purpose.

A detailed discussion of microwave methods for plasma investigations can be found in a review of Golant [66] and in several monographs [67, 68]. The resonator method requires measurements of the characteristics of a high-frequency resonator before and after formation of a plasma inside this resonator. The appearance of a plasma alters the resonance frequency and Q factor. Measurements of the resonance frequency shift and of the change in the reciprocal of the Q factor yield the active and reactive components of the complex plasma conductivity σ and the related electron density N_e as well as the effective frequency of collisions between electrons and molecules ν_{eff}.

When an alternating electric field $E = E_0 \exp(i\omega t)$ is applied to a plasma, the total density of the high-frequency current is governed by the directional part of the electron velocity distribution function of Eq. (1.4):

$$\mathbf{j} = e \int \mathbf{v} f dv = \frac{4\pi e}{3} \int_0^\infty \mathbf{f}_1 v^3 dv. \tag{2.36}$$

The solution of the transport equation for $\mathbf{f}_1$ under steady-state homogeneous conditions ignoring electron−electron interactions is [14]

$$\mathbf{f}_1 = -\frac{e}{m} \frac{\mathbf{E}}{\nu(v) + i\omega} \frac{\partial f_0}{\partial v}, \tag{2.37}$$

where $\nu(v)$ is the frequency of collisions between electrons and molecules, which depends on the electron velocity; f_0 is the symmetric part of the distribution function. Substituting Eq. (2.37) into Eq. (2.36), we obtain

$$\mathbf{j} = \frac{4\pi e^2 N_e}{3m} \mathbf{E} \int_0^\infty \frac{\nu - i\omega}{\nu^2 + \omega^2} \frac{\partial f_0}{\partial v} v^3 dv, \tag{2.38}$$

where N_e is the electron density.

On the other hand, it follows from the definition of the complex conductivity σ that

$$\mathbf{j} = \sigma\mathbf{E} = (\sigma_r + i\sigma_i)\mathbf{E}. \tag{2.39}$$

Comparing Eqs. (2.38) and (2.39) we can see that the real and imaginary parts of the complex conductivity can be expressed in terms of the derivative of the electron velocity distribution function:

$$\sigma_r = -\frac{4\pi e^2 N_e}{3m} \int\limits_0^\infty \frac{\nu}{\nu^2 + \omega^2} \frac{\partial f_0}{\partial v} v^3 dv, \tag{2.40}$$

$$\sigma_i = \frac{4\pi e^2 N_e}{3m} \int\limits_0^\infty \frac{\omega}{\nu^2 + \omega^2} \frac{df_0}{dv} v^3 dv. \tag{2.41}$$

It is usual to introduce the effective collision frequency defined by

$$\sigma_r = \frac{N_e e^2 \nu_{\text{eff}}}{m(\omega^2 + \nu_{\text{eff}}^2)}, \qquad \sigma_i = -\frac{N_e e^2 \omega}{m(\omega^2 + \nu_{\text{eff}}^2)}. \tag{2.42}$$

[The above expressions give the components of the conductivity for $\nu(v) = $ const, i.e., when the collision frequency is independent of the electron velocity.]

The real and imaginary parts of the conductivity can be determined using a high-frequency resonator. If a part of the resonator is filled with a plasma, the natural (resonance) frequency of the resonator ω_0 and its Q factor are affected. In the first approximation of the perturbation theory [69] the changes in these two resonator parameters are related to the complex conductivity of the plasma σ by

$$\frac{\Delta\omega}{\omega_0} = \frac{4\pi}{\omega_0} \int\limits_{V_1} \sigma E^2 dV \Big/ \int\limits_{V_2} E^2 dV, \tag{2.43}$$

where the integral in the numerator is taken over that part of the resonator volume which is filled with the plasma, whereas the integral in the denominator is taken over the whole resonator volume; E is the electric field for the selected oscillation mode. Substituting Eq. (2.42) into Eq. (2.43) and assuming that the collision frequency is independent of the coordinates, we obtain the following expressions for the frequency shift and the change in the reciprocal of the Q factor:

$$\frac{\Delta\omega}{\omega_0} = \frac{\bar{N}_e}{N_{\text{cr}}} \frac{A}{1 + \gamma^2}, \tag{2.44}$$

$$\Delta\left(\frac{1}{Q}\right) = \left(\frac{1}{Q} - \frac{1}{Q_0}\right) = 2\frac{\bar{N}_e}{N_{\text{cr}}} \frac{A\gamma}{1 + \gamma^2}, \tag{2.45}$$

where $\gamma = \nu_{\text{eff}}/\omega$; $N_{\text{cr}} = m\omega^2/4\pi e^2$ is the critical electron density at which the field frequency becomes equal to the plasma frequency; $\bar{N}_e$ is the volume-average electron density; $A = \int\limits_{V_1} E^2 dv \Big/ \int\limits_{V_2} E^2 dv$ is the form factor. The electron density and collision frequency can be found from Eqs. (2.44) and (2.45):

$$\bar{N}_e = \frac{N_{\text{cr}}}{A} \frac{\Delta\omega}{\omega} \left\{1 + \left[\frac{\Delta(1/Q)}{2\Delta\omega/\omega}\right]^2\right\}, \tag{2.46}$$

$$\nu_{\text{eff}} = \frac{\omega}{2} \frac{\Delta(1/Q)}{\Delta\omega/\omega}. \tag{2.47}$$

If $\omega^2 \gg \nu_{\text{eff}}^2$, the second term in the braces in Eq. (2.46) is small and the electron density is governed only by the resonance frequency shift.

The form factor A depends on the spatial distribution of electrons in the plasma. It can be found using graphs given in [66], where it is calculated for various modes in the case of a homogeneous distribution $N_e(r) = $ const and a diffuse one $N_e(r) = N_0 J_0(2.4r/a)$. The formula for the form factors in the case of TM_{0m0} modes in cylindrical resonators is given in [70]:

$$A = \frac{(r/R)^2}{2J_1^2(\chi_{0m})}\left[1 - \frac{1}{4}\left(\chi_{0m}\frac{r}{R}\right)^2 + \frac{1}{32}\left(\chi_{0m}\frac{r}{R}\right)^4 - \cdots\right], \qquad (2.48)$$

where r and R are, respectively, the radii of the discharge tube and the resonator; χ_{0m} is the m-th root of a zeroth-order Bessel function; J_1 is a first-order Bessel function.

However, these calculations of the form factors ignore the real distribution of the electric field in a resonator, which may differ from that calculated because of "sagging" of the field at the coupling apertures and because of the presence of a dielectric enclosure containing a plasma as well as of elements for coupling to the high-frequency channel. The resonator form factor can be determined also by an empirical method. It follows from the definition of the conductivity [66]

$$\jmath = \jmath_r + i\sigma_i = \jmath_r + i\frac{\varepsilon-1}{4\pi}\omega \qquad (2.49)$$

and from Eq. (2.42) subject to $\omega^2 \gg \nu^2$ that

$$\varepsilon = 1 - \frac{4\pi N_e e^2}{m\omega^2} = 1 - \frac{N_e}{N_{\mathrm{cr}}} = 1 - \frac{\omega_p^2}{\omega^2}, \qquad (2.50)$$

where ε is the permittivity of the plasma; $\omega_p^2 = 4\pi N_e e^2/m$ is the plasma frequency. Thus, the plasma permittivity is always less than unity and below the plasma frequency it is negative. It is usual to measure the electron density in a resonator at frequencies $\omega^2 \sim 0.1\omega_p^2$, i.e., at frequencies such that the plasma permittivity does not differ greatly from unity. Plasma can be modeled by a dielectric sample of permittivity close to unity (for example, foamed plastic) and in this way the resonator can be calibrated in respect of the frequency shift [71]. The form factor obtained in this way is 1.5-1.7 times smaller than that calculated from Eq. (2.48).

Apparatus for Determination of Electron Density by a

Resonator Method

A block diagram of the apparatus is shown in Fig. 9. Frequency- or amplitude-modulated microwaves generated by a klystron oscillator K were passed through a ferrite isolator I to a resonator R. We used standard-signal klystron oscillators of the G4-8, G4-9, and G4-10A type. The resonator was cylindrical and had different diameters (25, 16, 8 cm, etc.). There

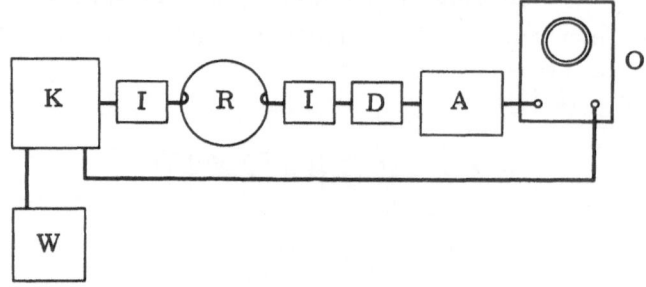

Fig. 9. Block diagram of the apparatus used in measurements of the electron density by the resonator method.

were apertures on the resonator axis for the insertion of a discharge tube. The oscillation modes excited in the resonator were M_{010}, TM_{020}, and TM_{030}.

The electric field of all these modes had only the axial (along the z axis) component, dependent on the radial coordinate, and it could be described by a zeroth-order Bessel function. The second index in the mode designation indicated how many times the electric field vanished along the resonator radius. The electric field maximum was located on the resonator axis where a tube with a plasma was placed; this increased the sensitivity of the electron density determination.

At high electron densities the perturbation theory was no longer valid and the dependence of the resonance frequency shift on the electron density was no longer linear. Considerable discrepancies for the TM_{0m0} oscillations began from electron densities $N_e \sim N_{cr}$ [71]. In our experiments the working frequency range $\omega = (2-4.6) \cdot 10^{10}$ sec^{-1} corresponded to critical electron densities $N_{cr} = (1.25-6.9) \cdot 10^{11}$ cm^{-3}.

In our configuration the resonator was used as a reentrant cavity. The resonator characteristics were deduced from the frequency dependence of the transmitted power. The amplitude-modulated output voltage from the resonator was detected in a detector D amplified by a wide-band amplifier A, and applied to the vertical plates of an oscillograph O. The horizontal scan of the oscillograph was provided by synchronizing pulses supplied by the klystron oscillator. We employed both frequency and amplitude modulation techniques.

In measurements of the resonance frequency shift it was more convenient to use the frequency modulation technique for small deviations of the frequency. A frequency-modulated signal was transformed by the resonator into one which was amplitude-modulated. The changes in the amplitude corresponding to the rising and falling branches of the resonance curve of the resonator were in antiphase. The amplitude modulation was minimal when the klystron oscillation frequency coincided with the natural frequency of the resonator. Thus, the moment of frequency coincidence could readily be determined by displaying two or three periods on the oscillograph screen.

The resonance curve was recorded to determine the Q factor of the resonator. In this case an amplitude-modulated signal was used. After detection and amplification, the signal was displayed on the oscillograph screen and measured or it was deduced from the output voltage of the amplifier. The oscillation frequency was measured with a heterodyne wavemeter W (of the ShGV-S type) or an internal wavemeter built into the klystron oscillator.

An electric discharge usually heated the tube so that there was an additional resonance frequency shift and the width of the resonance curve changed because the dielectric properties of the tube material were temperature-dependent. In the case of quartz the dielectric losses were practically independent of temperature. However, the losses in glasses of various kinds with different admixtures increased by a factor of 2-3 when the temperature was raised from 20 to 250°C [72].

In the first experiments this additional shift was allowed for on the assumption that it was proportional to the tube temperature and that the degree of heating of a glass tube was proportional to the square of the discharge current. In subsequent experiments we placed a copper—constantan thermocouple on the outer surface of the tube near the resonator and the readings of this thermocouple were used to correct the resonance frequency shift and resonance broadening to allow for the heating of the glass.

Errors in Measurement of Electron Density and

Collision Frequency

The minimum frequency shift and the corresponding minimum electron density which can be measured by the resonator method are governed by the Q factor of the resonator and by the change in this factor because of introduction of a plasma [67]:

$$\left(\frac{\Delta\omega}{\omega}\right)_{min} = c\left[\frac{1}{Q} + \Delta\left(\frac{1}{Q}\right)\right], \qquad (2.51)$$

where c is a coefficient of the order of 0.25. The Q factor of the empty resonator (R = 8 cm, r = 1.2 cm, h = 3 cm, oscillation mode TM_{020}) was about 3000. Introduction of a glass tube reduced this factor to 2000. Changes in the reciprocal of the Q factor depended on the collision frequency, i.e., on the gas pressure. For example, at a pressure of p = 1 Torr the change was $\Delta(1/Q) = 0.5 \cdot 10^{-3}$, whereas at p = 5 Torr, the change was $\Delta(1/Q) \simeq 2 \cdot 10^{-3}$. Substituting Eq. (2.51) into Eq. (2.46), we readily obtained the minimum electron density which could be determined. For p = 1 Torr, we found that $N_{emin} = 0.6 \cdot 10^9$ cm^{-3} whereas at p = 5 Torr the corresponding value was $N_{emin} = 1.6 \cdot 10^9$ cm^{-3}.

The precision of the determination of the electron density depended also on the method used to measure the frequency and on the accuracy of the form factor A. Allowance for all these points indicated that the error in the electron density determination was about 15-20%. At higher pressures the broadening of the resonance curve was greater and the precision of the determination of the frequency shift decreased so that the error could reach 30-35%.

The effective collision frequency was deduced from Eq. (2.47) subject to an error much larger than in the electron density. An estimate of the relative error gave 30-40%.

§ 3. Construction of Discharge Tube,

Probes, and Power Supply System

Discharges took place in glass or quartz tubes. The easiest to make were molybdenum glass tubes. However, molybdenum glass suffered from dielectric losses larger than those in Pyrex glass and much larger than in quartz. The most suitable tube material was fused quartz (silica). The loss-angle tangent of fused quartz was 16 times lower than that of ordinary glass [73].

A discharge tube used in plasma parameter measurements and in laser experiments is shown schematically in Fig. 10. It had a water jacket for cooling of the gas inside the tube. The electrodes were made of molybdenum. The water jacket had a central dip (5 cm long) for a cylindrical resonator. The resonator used in our study was cut along its diameter so that it could be mounted easily around the discharge tube. The tube length was 40-100 cm. A system of electrical probes, introduced into the tube by a ground-glass joint, was placed on the anode side of the tube. Cylindrical probes were used in the determination of the electron energy distribution function. The main results were obtained using probes of 0.04 mm diameter and about 2-4 mm long. The probes were insulated by glass capillaries of about 0.1 mm diameter. The longitudinal field was relatively high, of the order of 20 V/cm or 2 V/mm. Therefore, a cylindrical probe was oriented at right angles to the tube axis to ensure that all parts of the probe were, if possible, located on one equipotential surface. Since our measurements indicated that the ratio of the directional (drift) and random electron velocities was

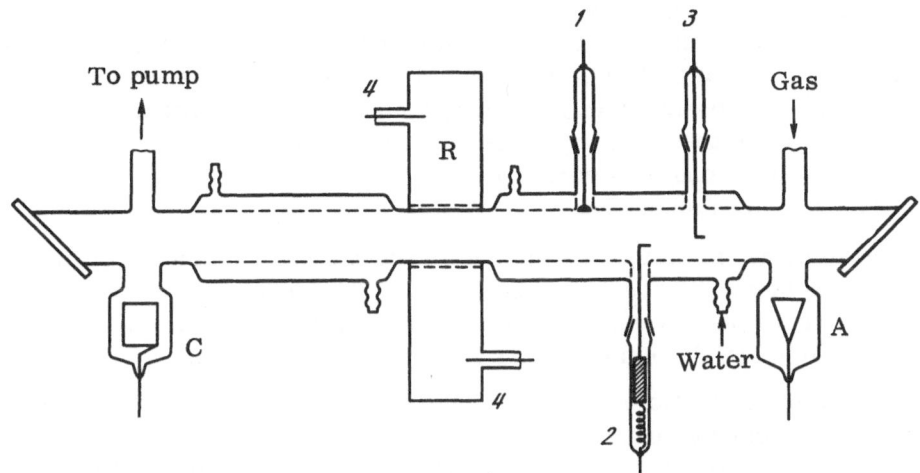

Fig. 10. Schematic diagram showing a discharge tube: A is the anode;
C is the cathode; R is the resonator; 1-3 are probes and 4 are pins.

small ($\delta_{eff} \sim 10^{-2}$), the electron distribution could be regarded as almost isotropic. Therefore, the characteristics recorded with a perpendicular probe should not be distorted by the directional motion of electrons. However, the precision of the determination of the vacuum potential was then somewhat less.

We used molybdenum, gold, and platinum probes. There were no significant differences between the results obtained using probes made of different materials. Figure 10 shows a planar (flat) wall probe 1, a probe which could be moved along the radius 2, and a simple axial probe 3.

The resonator was excited by pins 4, built into the end wall of the resonator. The degree of coupling could be varied by micrometer screws.

The ends of the tube were closed by NaCl plates oriented at the Brewster angle; this was necessary to obtain stimulated emission.

A discharge tube was supplied from a high-voltage rectifier operating in a double-half-period circuit. This rectifier could supply voltages up to 10 kV and currents up to 100 mA. A special feature of this rectifier was a smoothing filter with a 150 μF capacitance. Consequently, the pulsations of the rectifier output did not exceed 0.02%, which helped in the probe measurements. The ballast resistance could be varied from 20 to 100 kΩ.

A vacuum system reduced the pressure in the tube to p = 10^{-3} Torr. The gas-admission system could be used for specially prepared mixtures and pure gases. The gas velocity was about 1 m/sec.

CHAPTER III

RESULTS OF INVESTIGATIONS OF GAS-DISCHARGE PLASMAS IN MOLECULAR GASES

Since a probe located in a plasma perturbs the surrounding region, the range of conditions under which the probe method can be used is limited. The degree of perturbation caused by a probe is governed by the ratio of the probe radius to the mean free path of electrons in the plasma a/λ_e. The theory [38] demands that this ratio be small and this imposes an upper limit to the range of pressures in which the probe method can be used. If we assume that the probe radius a is the smallest realistic value of 0.02 mm and if we assume that the ratio $a/\lambda_e = 0.1$ is satisfactory, a pressure of 2 Torr is the upper limit in probe measurements. In molecular gas discharges the gas temperature is 1.5-3 times higher than the room value [74, 75]. Expulsion of the hot gas from the discharge tube reduces the number of molecules and increases the mean free path of electrons by a corresponding factor. Consequently, we can increase the upper limit of the pressures somewhat.

Bearing these points in mind, we confined the probe measurements of the distribution function to pressures below 4 Torr. The lower limit of the range of pressures (p = 1 Torr) in which the probe measurements were carried out was governed by the appearance of standing losses in discharge tubes of 20 mm diameter in almost all mixtures.

Measurements of the collision frequency and density of electrons by the high-frequency resonator method were carried out between 0.5 and 6 Torr. It was difficult to use the resonator method at pressures above 6 Torr because of the strong broadening of the resonance profile, i.e., because of a reduction in the Q factor of the resonator.

§ 1. Investigations of Positive Discharge Columns in Nitrogen

Discharges in pure nitrogen do not yield the information related directly to the CO_2 laser operation. However, we investigated positive columns of glow discharges in nitrogen for the following reasons. First of all, discharges in a pure gas were simpler to analyze than those in a mixture of chemically reacting gases, such as those used in carbon dioxide lasers. Since the dissociation energy of the nitrogen molecules was fairly high, $\varepsilon_d = 9.7$ eV, the degree of dissociation of pure nitrogen was slight [76]. The main ions in nitrogen discharges, as established by mass spectrometry [77], were the N_2^+ ions. Nevertheless, we should remember that nitrogen has, perhaps more than any other molecule, a rich spectrum of various energy states. Therefore, investigations of electrical characteristics of discharges in pure nitrogen make it possible to determine the main properties of molecular plasma without the complicating influence of the dissociation processes and chemical reactions.

Secondly, we carried out this study because of the availability of extensive data on the transport coefficients (drift velocities, electron and ion mobilities, and diffusion coefficients) for nitrogen under similar conditions to those in our experiments; moreover, measurements [6] and calculations [58] have been made of the electron energy distributions. Naturally, it would be interesting to compare the results of our measurements with those in the literature.

Electron Energy Distribution

It is usual to give the results of measurements of the electron energy distribution in the form of the function f_0(eV) of Eq. (2.3) or, which is equivalent, as the second derivative of the probe current with respect to the voltage, multiplied by the square root of the potential. It seems to be more convenient and correct to give the logarithm of the second derivative. Firstly, the representation on a semilogarithmic scale makes it possible to cover a wider range of the second derivative. Secondly, it is clear from Eq. (2.2) that the second derivative is proportional to the symmetric part of the distribution function. This means that the classical distributions can be represented by simple graphs on this scale. For example, the Maxwellian function (1.8) is a straight line, the Druyvesteyn function (1.16) is a parabola, etc. Naturally, a comparison of the experimental and theoretical curves is then simple and clear. Thirdly, the results of calculations of the theoretical distribution functions are then given by dependences $\log f_0(\varepsilon)$, which also facilitates a comparison of the curves.

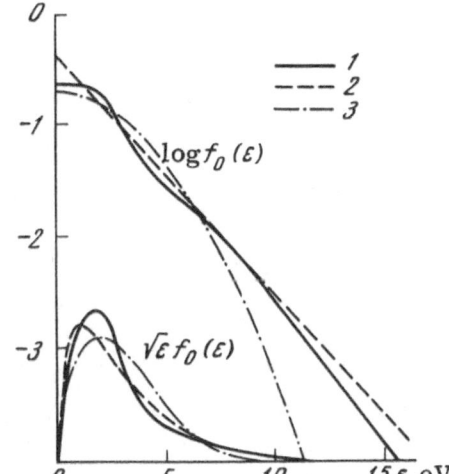

Fig. 11. Electron energy distribution in pure nitrogen: 1) experimental (f_0^{exp}); 2) Maxwellian (f_0^M); 3) Druyvesteyn (f_0^D).

Figure 11 shows, on a logarithmic scale, the electron energy distribution function $f_0(\varepsilon)$ of pure nitrogen at p = 1 Torr for a discharge current i_d = 42 mA in a tube of 20 mm diameter. This curve is normalized in such a way that

$$\int_0^{\varepsilon_k} \sqrt{\varepsilon}\, f_0(\varepsilon)\, d\varepsilon = 1. \tag{3.1}$$

The average energy in this distribution, deduced from the formula

$$\bar{\varepsilon} = \int_0^{\varepsilon_k} \varepsilon^{3/2} f_0(\varepsilon)\, d\varepsilon, \tag{3.2}$$

is 3.02 eV. We can see from Fig. 11 that the distribution function does not fall to zero at ε = 0. This is a consequence of the analysis of the second derivative on the basis of Eq. (2.33).

Figure 11 includes, for the sake of comparison, the Maxwellian and Druyvesteyn distribution functions for the same average energy. Neither of these classical distributions fits the observed curve although at low electron energies (ε < 3 eV) the distribution function is close to the Druyvesteyn curve, whereas at energies ε > 6 eV it is close to the Maxwellian curve. A characteristic feature of the experimental curve is a fall in the number of electrons in the range ε = 3-5 eV. The lower part of Fig. 11 shows, for the sake of comparison, the same functions on a linear scale but multiplied by $\sqrt{\varepsilon}$. The most probable electron energies are different for the experimental and theoretical distributions.

The distribution of electron energies in nitrogen is practically unaffected when the discharge current is varied. The second derivative of the probe current rises in amplitude with the current without a change in shape. This is demonstrated in Fig. 12, which shows the distribution functions for p = 1 Torr and different discharge currents i_d = 21, 42, and 61 mA. It is clear from Fig. 12 that the second derivatives terminate at electron energies ε = 10-13 eV. The number of electrons participating in the excitation of high-energy states and in the ionization process is two or three orders of magnitude less than the number of electrons with the average energy. The amplitude of the second derivative at these high energies becomes very low and comparable with the noise background. Additional information on the behavior of the distribution function at high energies was obtained by us from spectral measurements.

As shown in [78], the excitation of the $C^3\Pi_u$ electronic state of the N_2 molecule takes place as a result of electron impact from the ground state $X^1\Sigma_g^+$. The cross section of this

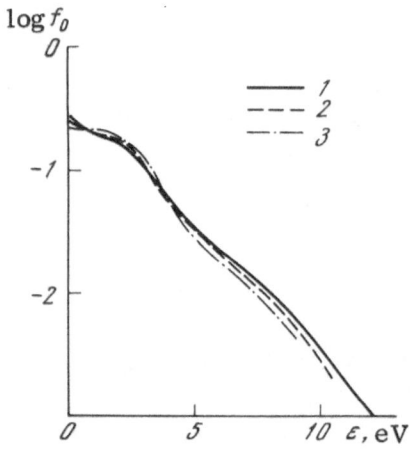

Fig. 12. Electron energy distributions in N_2 obtained for different discharge currents i_d (mA): 1) 61; 2) 42; 3) 21.

process is fairly large and reaches 10^{-16} cm^2. The vibrational levels of the C$^3\Pi_u$ state are deactivated exclusively by the emission of the second positive system. The average lifetime of the C$^3\Pi_u$ state is $\tau \sim 10^{-8}$ sec [79], which is much less than the time constants of the vibrational and rotational thermalization and diffusion to the walls under our discharge conditions. Therefore, changes in the intensities of the 2$^+$ bands of the nitrogen system are due to changes in the distribution function in the range of electron energies equal to the excitation thresholds of these bands.

Figure 13 shows dependences of the intensity of the (0, 2) band of the 2$^+$ nitrogen system on the discharge current (10–50 mA) obtained at two pressures of 1.4 Torr (curve 1) and 2.7 Torr (curve 2). The other bands of the 2$^+$ system behave similarly. The dependences of the band intensities on the discharge current are practically linear. Some slowing down of the rise of the intensity at the high discharge currents is probably due to the fact that the gas temperature rises and the gas density decreases with increasing discharge current. (The rise of the gas temperature in the discharge altered the distribution of intensities in the rotational band structure. Therefore, in spectral measurements we used the total band intensities and not the intensities of the edges or of individual rotational lines.)

The dependences of the electron energy distributions on the pressure in a discharge are illustrated in Fig. 14, which shows the distribution functions obtained at pressures of 1 and 2 Torr using discharge currents of 21 and 25 mA, respectively. A comparison of these curves shows that at high pressures the "tail" of the distribution falls somewhat faster. The average

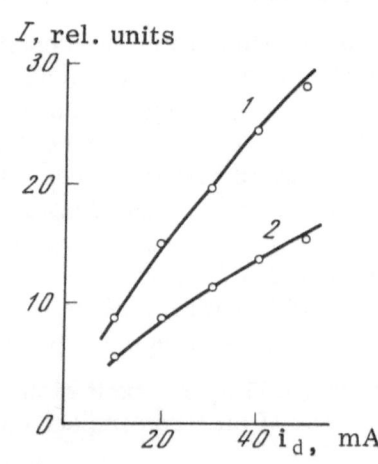

Fig. 13. Dependences of the (0, 2) band intensity of the 2$^+$ system of N_2 on the discharge current in N_2.

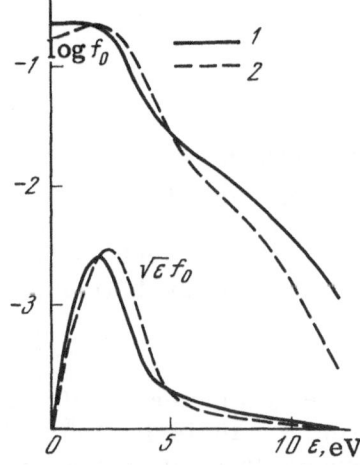

Fig. 14. Electron energy distributions in N_2 discharges: 1) p = 1 Torr, $\bar{\varepsilon}$ = 3.1 eV; 2) p = 2 Torr, 2.8 eV.

electron energy decreases by 10-15%. This behavior of the energy distribution is confirmed by the spectral measurements. When the pressure is increased, the intensity of the 2^+ bands falls strongly. For example, when the pressure is altered from 1 to 2 Torr, the intensity falls by a factor exceeding 1.5. Similar changes in the band intensities, agreeing qualitatively with our results, were obtained under analogous discharge conditions by other workers (the pressure dependences were studied in [80] and the current dependences in [81]).

Density of Electrons and Frequency of Their Collisions with Molecules

Figure 15 shows the dependences of the electron density on the discharge current at three pressures p = 1, 2, and 3 Torr (curves 1-3, respectively) measured by the microwave resonator method. We recall that the electron density given by this method is the average over the discharge-tube cross section. It is clear from Fig. 15 that the electron density varies linearly with the current between 10 and 60 mA in a tube of 20 mm diameter. This behavior follows from measurements of the probe current at the vacuum potential, from the areas under the distribution curves $f_0'(\varepsilon)$, and from the intensities of the 2^+ bands of N_2.

When the gas pressure is increased, the electron density rises somewhat: For example, when p is increased from 1 to 3 Torr, $\bar{N}_e$ rises by 25-30%.

The frequency of collisions between electrons and molecules is also a function of pressure (Fig. 16). It is clear from this figure that the collision frequency is not directly propor-

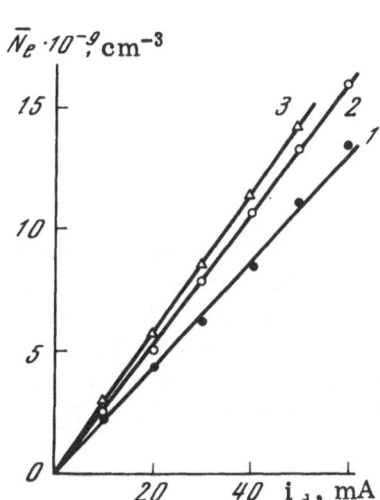

Fig. 15. Dependences of the electron density in N_2 discharges on the discharge current i_d.

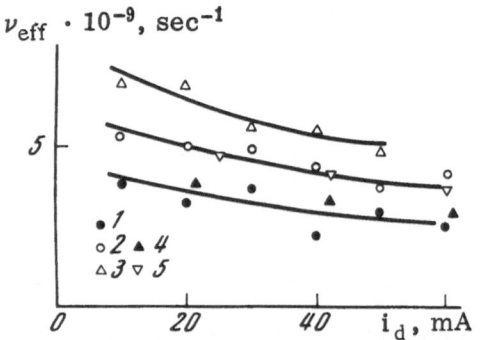

$\nu_{eff} \cdot 10^{-9}, \; sec^{-1}$

Fig. 16. Effective frequency of collisions between electrons and N_2 molecules measured by the microwave method (p = 1, 2, 3 Torr for curves 1–3, respectively) and calculated using Eq. (3.9) with p = 1 Torr (4) and 2 Torr (5).

tional to the pressure. For example, an increase in the pressure by a factor of 2 (from 1 to 2 Torr) corresponds to an increase in the collision frequency from $3 \cdot 10^9$ to $5 \cdot 10^9 \; sec^{-1}$ when the current is 30 mA. This is due to an increase in the gas temperature with rising pressure and the reduction in the number of neutral praticles in the discharge gap. If we use the temperatures given in [75], we can readily obtain the reduced collision frequency ν_{eff}/N. It is then found that this reduced frequency is approximately the same at different pressures. Some fall in the collision frequency with rising current is also due to the rise of the gas temperature.

Electric Field Intensity

The electric field was measured with pairs of probes located on the tube axis. An S–50 electrostatic voltmeter was connected to these probes and the voltage drop in the column between the probes was measured. Since the current in the probe circuit was zero, the probe potentials were equal to the floating potential. If the properties of the plasma and, particularly, the electron energy distributions were the same at the points of location of the probes, the measured potential differences should be equal to the differences between the plasma potentials. Control experiments involving simultaneous measurements by the electrostatic voltmeter and by the widely used compensation method [38] confirmed the validity of this hypothesis.

Knowing the distance between the probes we could readily obtain the electric field in the positive column. Figure 17 shows the dependences of the electric field on the current flowing

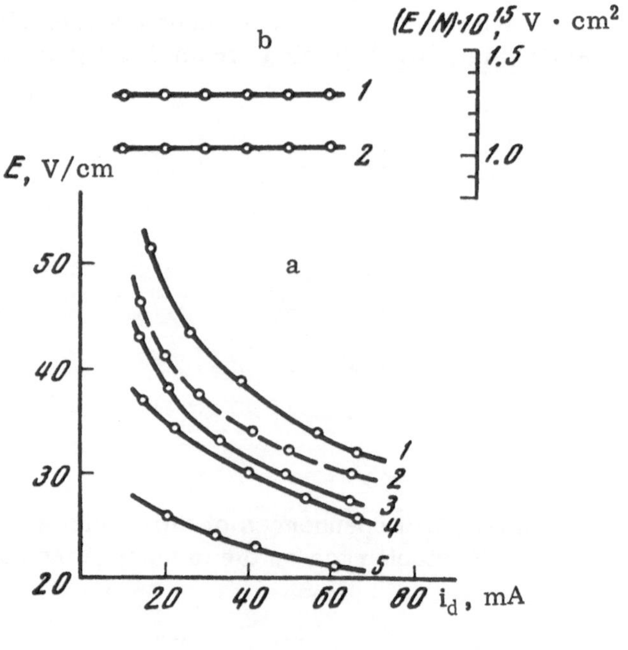

Fig. 17. Dependences of the longitudinal electric field E (a) and of the reduced field E/N (b) on the discharge current i_d in a tube of 20 mm diameter. a: 1) N_2, p = 4 Torr; 2) $CO_2 - N_2$ (1:1), p = 2.7 Torr; 3) N_2, p = 2.6 Torr; 4) N_2, p = 2.0 Torr; 5) N_2, p = 1.0 Torr. b: 1) N_2, p = 1 Torr; 2) N_2, p = 2 Torr.

in a nitrogen-filled tube of 20 mm diameter. The field intensity decreased with the rising current, i.e., the current–voltage characteristics were of the following type. The field intensities in the positive column were quite high: 20-50 V/cm. The falling nature of the current–voltage characteristics could be due, as in the case of the band intensities, to the rise in the gas temperature with increasing discharge current.

The electron gas energy can be described by the reduced field E/N. Figure 17 included also the dependence of E/N, which is found to be practically independent of the current. The number of molecules per unit volume N used in this case was calculated taking the gas temperatures from [75].

Discussion of Results

Distribution Function. The degree of ionization in our experiments is low, $N_e/N \sim 10^{-7}$, and estimates of the collision frequencies obtained using Eqs. (1.7) and (1.10) indicate that we are dealing with a weakly ionized plasma satisfying the inequality $\nu_e \ll \delta_{eff} \nu_m$. We can then expect a Druyvesteyn distribution (1.16) for the electron energies. If we use the formula (1.17) for the average electron energy in the Druyvesteyn distribution and calculate it for the field intensities and mean free paths in our experiments, we find that $\bar{\varepsilon} \sim 30\text{-}50$ eV, i.e., the calculated value is more than an order of magnitude greater than the values found experimentally. This enormous discrepancy is due to the important role played by inelastic processes in molecular plasmas, which are ignored in the derivation of the Druyvesteyn distribution.

Electrons in molecular plasmas lose their energy in the excitation of rotational, vibrational, and electronic degrees of freedom and in the dissociation of molecules. Since the electron energy distribution function is influenced by inelastic and elastic collisions with molecules, the similarity of the experimental distributions reported earlier is purely formal and has no ontological meaning.

One of the main inelastic processes in the scattering of electrons by the N_2 molecules is the excitation of the vibrational levels in the ground electronic state. The measurements of Schulz [5] indicated that the excitation cross sections of the first eight vibrational levels are very large (due to resonance) reaching maximum values (at $\varepsilon = 2.5$ eV) amounting to $\sigma_{max} = 3 \cdot 10^{-16}$ cm². All the experimental electronic energy distribution curves have a sharp fall beginning at 2.5-3 eV. This fall is due to the vibrational excitation of the N_2 molecules. The experimental distribution function is compared in Fig. 18 with those calculated from the tran-

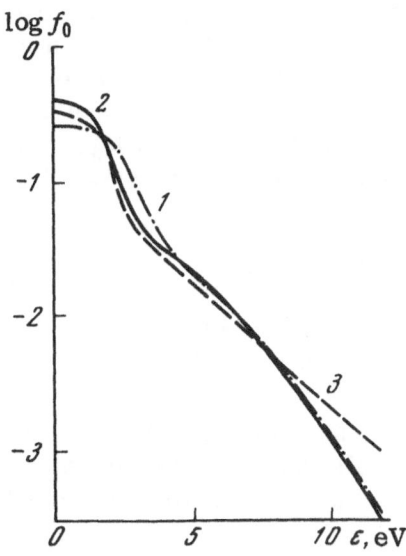

Fig. 18. Electron energy distributions in N_2: 1) experimental curves; 2) calculation [58]; 3) calculation [82]; $E/N = 10^{-15}$ V·cm².

sport equation by Lukovnikov et al. [58] and by Nighan [82] for the same reduced field E/N = 10^{-15} V · cm^2. The curves calculated in [58] and [82] are in good mutual agreement in the energy range up to 8-9 eV. They diverge at $\varepsilon > 9$ eV because of the use of different excitation cross sections of the electronic state B$^3\Pi_g$: The experimental cross section taken from [83] was used by Lukovnikov et al. [58], whereas Nighan [82] employed an unidentified cross section with a threshold at $\varepsilon_i = 14.5$ eV. Since in this range of energies our experimental function agreed better with that of Lukovnikov et al. [58], we assumed that the cross sections selected in [58] were more reliable than the cross sections [18] used in [82].

In the energy range 2-4 eV there are some discrepancies between our experimental curve and both calculated curves. These discrepancies are clearly due to the fact that collisions of the second kind between excited molecules and electrons are ignored in the calculations. Since the N$_2$ molecule contains identical nuclei, optical transitions between the vibrational levels in any one given electronic state are forbidden. The vibrationally excited N$_2$ molecules can lose their energy either by interaction with the tube walls or as a result of collisions of the second kind with electrons. This, in conjunction with large vibrational excitation cross sections, should make for high vibrational temperatures in the discharge. In fact, as shown by us in [78], the vibrational temperature of pure nitrogen can reach 4000°K and it falls on addition of CO$_2$, whose molecules can deexcite nitrogen.

When the vibrational temperature is 4000°K, about 70% of the N$_2$ molecules are in the excited state. If we assume that the molecular diffusion coefficient at p = 1 Torr is of the order of 10^2 cm^2/sec, we find that the diffusion lifetime of molecules in a tube of 2 cm diameter is $\tau_{diff} \sim R^2/D \sim 10^{-2}$ sec. The lifetime of the excited molecules deactivated by electron impact can be estimated from the principle of detailed equilibrium: $\tau_{deact} = \exp(-AE/kT_{eff})/N_e \langle \sigma_{vib} v \rangle \sim 3 \times 10^{-1}$ sec (at p = 1 Torr and for $i_d = 30$ mA). The application of the principle of detailed equilibrium presupposes an equilibrium distribution of electrons between the energies, which is not true; however, this can be done in order-of-magnitude estimates. It follows from the quoted values that although the diffusion lifetime is shorter than the deactivation lifetime, these quantities are of the same order of magnitude. We must bear in mind that when the discharge current and pressure are increased, the difference between τ_{diff} and τ_{deact} decreases. Clearly, allowance for collisions of the second kind between vibrationally excited molecules and electrons should result in some broadening of the dip in the distribution function due to vibrational excitation and should reduce the discrepancy between the calculated and experimental curves.

This discrepancy may also be due to other processes involving highly excited electronic states of the N$_2$ molecule, which are also ignored in the calculations. The nitrogen molecule has a metastable singlet state $a^1\Pi_g$ [84] whose energy is about 8.25 eV. This state participates in the "active nitrogen" mechanism [85]. As is known, the spectrum of the afterglow of active nitrogen includes prominently the first positive system of nitrogen (B$^3\Pi_g$ − A$^3\Sigma_u^+$). According to [85], the B$^3\Pi_g$ state may be populated with the aid of the $a^1\Pi_g$ state. The relatively high electron density in the afterglow ($\sim 10^9$ cm^{-3}) favors this process:

$$N_2(a^1\Pi_g) + e \ (\text{slow}) \rightarrow N_2(B^3\Pi_g) + e. \tag{3.3}$$

The electron temperature was measured in [86] by the probe method in the afterglow of nitrogen.* Some increase in the electron temperature in a period of 5 msec was observed. Clearly,

---

* In the light of our results there are some doubts about the Maxwellian distribution of electrons assumed in [86]. Therefore, one should view critically the absolute values of the electron temperatures given in that paper. However, the qualitative behavior of the electron energy is given correctly (see § 5).

the process (3.3) was responsible for this increase. Moreover, the same process could be important also in static discharges and it could cause electron heating.

Average Energy of Electrons. Since the electron energy distribution in our plasmas is not Maxwellian, it is meaningless to speak of the electron temperature in its usual sense. In this case the electron energy can be described by the average value $\bar{\varepsilon}$ given by Eq. (1.3). For a Maxwellian distribution we have $\bar{\varepsilon} = (3/2)kT_e$, where T_e is the electron temperature. Using this expression, some workers speak of the effective electron temperature in a non-Maxwellian plasma, i.e.,

$$kT_{\text{eff}} = {}^2/_3 \bar{\varepsilon}. \tag{3.4}$$

The average energy is a very important characteristic of the electron component of the plasma; it would be interesting to compare the results obtained by us with those reported by other workers. Figure 19 gives the values of the average energy obtained in various investigations; these are plotted as a function of the reduced field E/N between $0.6 \cdot 10^{-15}$ and $2.0 \cdot 10^{-15}$ V·cm². A characteristic feature of all the values, including ours, is the rise of the average energy with increasing E/N. However, there are some differences between the absolute values.

The dashed curve Fig. 19 is taken from [18] and it is drawn along the experimental points of the characteristic energy $\varepsilon_k = eD_e/\mu_e$ taken from [87]. By definition,

$$D_e = \frac{1}{3}\lambda_e \bar{v}_e = \sqrt{\frac{2}{m}}\frac{1}{3N}\int_0^\infty f_0(\varepsilon)\frac{\varepsilon\,d\varepsilon}{\sigma_y(\varepsilon)}, \tag{3.5}$$

$$\mu_e = \frac{\bar{v}_d}{E} = -\sqrt{\frac{2}{m}}\frac{e}{3N}\int_0^\infty \frac{\varepsilon}{\sigma_y(\varepsilon)}\frac{df_0}{d\varepsilon}d\varepsilon, \tag{3.6}$$

where D_e and μ_e are the diffusion coefficient and the mobility of electrons, $\bar{v}_e$ and $\bar{v}_d$ are the average and drift velocities of electrons, $\sigma_y(\varepsilon)$ is the elastic scattering cross section, and $f_0(\varepsilon)$ is the symmetric part of the distribution function. In the case of a Maxwellian velocity distribution the quantity ε_k is the electron temperature expressed in volts and the average electron energy is then $1.5\varepsilon_k$. In the absence of a Maxwellian distribution the coefficient of proportionality between $\bar{\varepsilon}_k$ and $\bar{\varepsilon}$ is less than 1.5. Thus, the values of $\bar{\varepsilon}$ taken from [18] and recalculated with the aid of Eqs. (3.5) and (3.6) should lie between our values and those calculated from [58]. This group of experimental points differs from the average energies taken from [58, 82], which are 10-20% lower than our experimental values. One of the possible reasons for this discrepancy may be, as pointed out earlier, the lack of allowance for colli-

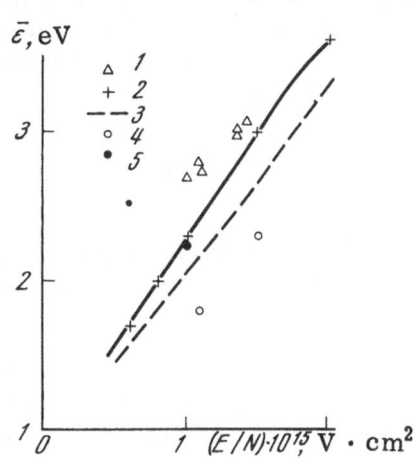

Fig. 19. Average energy of electrons in N_2: 1) experimental results; 2) taken from [58]; 3) taken from [18]; 4) taken from [88]; 5) taken from [82].

sions of the second kind between electrons and excited nitrogen molecules. These molecules may be vibrationally excited or they may be in highly excited metastable states. This hypothesis is supported by the observation that the discrepancy increases somewhat with decreasing value of the parameter E/N. This corresponds to higher pressures or tube diameters, i.e., to conditions under which the role of all possible metastable states is considerably greater.

Effective Fraction of Energy. Knowledge of the average electron energy and effective collision frequency allows us to find another important quantity $\delta_{eff}(T_{eff})$, which is the effective fraction of the energy of an electron lost in one collision with a molecule. It is very difficult to calculate directly this quantity [13] because of the need to know the cross sections of many inelastic processes but the application of the balance equation (1.9) simplifies the problem. In the steady-state case Eq. (1.9) can be expressed in the form

$$\delta_{eff}\bar{\varepsilon}\nu_{eff} = eE\bar{v}_d, \tag{3.7}$$

where $\bar{v}_d$ is the electron drift velocity, found from the equation for the current density

$$j = N_e e\bar{v}_d. \tag{3.8}$$

The experimental value of δ_{eff} is $2.2 \cdot 10^{-2}$ and $1.9 \cdot 10^{-2}$ when the average energy is 3 and 2.7 eV, respectively. We shall mention, for the sake of comparison, that the fraction of the electron energy lost only in elastic collisions is $\delta_e = (2m/M)_{N_2} = 0.4 \cdot 10^{-4}$, i.e., it is more than two orders of magnitude smaller than δ_{eff}.

Collision Frequency. If we know the electron distribution function, we can calculate the effective collision frequency from

$$\nu_{eff} = \frac{4\pi}{3} \int\limits_0^\infty \nu(v)\, v^3\, \frac{df_0}{dv}\, dv, \tag{3.9}$$

where $\nu(v) = Nv\sigma(v)$; $\sigma(v)$ is the total collision cross section. We calculated the value of ν_{eff}/N for each experimentally determined electron distribution curve. The points calculated from Eq. (3.9) were plotted in Fig. 16 for p = 1 and 2 Torr; we found that the values of ν_{eff} calculated from the distribution function were in good agreement with those found by the high-frequency method. The collision frequencies needed in the calculation of ν_{eff} were taken from [47, 48] and the values of the gas temperatures T_g needed in the calculation of $N = 3.53 \cdot 10^{16} \cdot 273p/T_g$ were taken from [75].

Ionization in the Positive Column of a Glow Discharge in N_2

A steady-state positive column in a glow discharge can be described by the Schottky theory [89] provided $\lambda_e \ll R$, where R is the column radius. According to this theory a gas is ionized by collisions between fast electrons and gas molecules in the ground state. Charged particles disappear as a result of ambipolar diffusion. Consequently, electrons and ions move along radial directions at the same velocities to the walls. If they do not recombine in the bulk of the tube but are neutralized on the wall, the balance equation for the charged particles can be expressed in the form

$$\tau_D z = 1, \tag{3.10}$$

where $\tau_D = 1/\mu^2 D_{am}$ is the diffusion lifetime of the particles; $\mu = 2.4/R$; z is the ionization frequency, i.e., the number of ionizing collisions between electrons and gas molecules per unit time.

Equation (3.10) allows us to determine experimentally the average lifetime of free electrons and ions in a plasma. We can do this by measuring the density of the ion current reaching the tube wall j_w and finding the number of free electron and ions per unit length of the col-

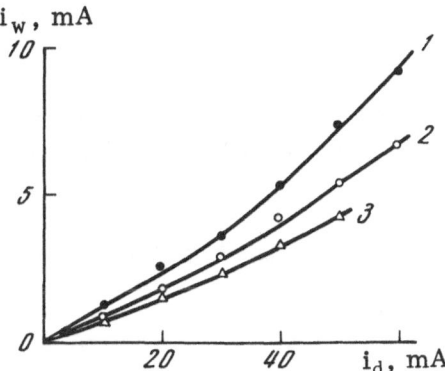

Fig. 20. Ion current flowing to a wall probe (probe diameter d_p = 0.3 cm) in N_2 discharges occurring in a tube of 20 mm diameter at p = 1, 2, and 3 Torr (curves 1-3, respectively).

umn $N_e \pi R^2$. In fact, if the tube radius is R, the number of ions leaving a positive column section 1 cm long in 1 sec is

$$j_w 2\pi R/e. \tag{3.11}$$

This number should be equal to the number of ions generated as a result of ionization of the gas in the same section of the positive column, i.e., it should be equal to $zN_e \pi R^2$. Thus,

$$\tau_D = 1/z = e\pi R^2 N_e / 2\pi R j_w. \tag{3.12}$$

Figure 20 gives the dependence of the ion current reaching the tube wall, measured with a flat probe of 0.3 mm diameter. It is clear from this figure that the ion current density rises superlinearly with the discharge current. When the pressure is increased, the density of the ion current to the wall decreases but the nonlinearity of the dependence on the discharge current is retained. We calculated the ionization frequencies by analyzing the results obtained in this way with Eq. (3.12) and using the electron densities plotted in Fig. 15. The results obtained were collected in Table 1. When the discharge current was increased from 10 to 60 mA, the ionization frequency rose by 30%. The absolute values of z_{exc} were within the range $(4-12) \cdot 10^4$ sec^{-1}. These values corresponded to a diffusion lifetime of the order of 10^{-15} sec. It was interesting to compare this value with the lifetime of ions governed by bulk recombination. If the dissociative recombination coefficient $\beta \sim 10^{-7}$ cm/sec was taken from [90], the value of $\tau_{rec} = 1/\beta N_e$ was found to be of the order of 10^{-3} sec, which was much longer than the diffusion lifetime of ions.

If the electron energy distribution function is known, the ionization frequency can be calculated as follows. The probability that an electron moving at a velocity v experiences an

TABLE 1

Calculated and experimental quantities	i_d, mA											
	10	20	30	40	50	60	10	20	30	40	50	60
	p = 1 Torr						p = 2 Torr					
i_w, μA	1.2	2.5	3.6	5.3	7.3	9.2	0.8	1.7	2.8	4.2	5.3	6.7
$j_w \cdot 10^2$, mA/cm^2	1.7	3.5	5.1	7.5	10.3	13.0	1.1	2.4	4.0	5.9	7.5	9.5
$N_e \cdot 10^{-9}$, cm^{-3}	2.2	4.3	6.2	8.4	11.0	13.4	2.5	5.0	7.8	10.6	13.2	15.9
$z_{exc} \cdot 10^{-4}$, sec^{-1}	9.5	10.3	11.2	11.2	11.7	12.2	5.6	6.0	6.5	7.0	7.1	7.5
$\langle \sigma_i v \rangle_{exc} \cdot 10^{12}$, cm^3/sec	4.2	5.2	5.7	6.6	7.2	7.9	1.5	1.8	2.1	2.5	2.7	2.9
$\langle \sigma_i v \rangle_{calc} \cdot 10^{12}$, cm^3/sec	0.40						0.12					
T_g, °K	430	485	530	565	590	630	510	590	635	680	710	750
$N \cdot 10^{-16}$, cm^{-3}	2.2	2.0	1.8	1.7	1.6	1.5	3.8	3.3	3.0	2.8	2.7	2.6
E, V/cm	29.0	25.9	24.1	22.8	21.7	21.0	39.0	35.2	32.0	30.0	28.1	26.6
$E/N \cdot 10^{15}$, V·cm^2	1.30	1.30	1.31	1.32	1.33	1.35	1.03	1.07	1.05	1.05	1.03	1.03

ionizing collision in a time interval dt is

$$N\sigma_i(v)\,vdt, \tag{3.13}$$

where $\sigma_i(v)$ is the ionization cross section. Electrons moving at a velocity between v and v + dv in 1 cm³ perform, in a time dt,

$$N\sigma_i(v)\,dt\,N_e\,4\pi v^2 f_0(v)\,dv \tag{3.14}$$

such collisions. The total number of ionization events in a time dt in 1 cm³ of the gas is

$$zN_e dt = \int_{v\,min}^{\infty} N\sigma_i(v)\,vdt\,N_e\,4\pi v^2 f_0(v)\,dv, \tag{3.15}$$

and hence

$$z = N\int_{v\,min}^{\infty}\sigma_i(v)\,v4\pi v^2 f_0(v)\,dv = N\langle \sigma_i v\rangle. \tag{3.16}$$

Since the experimentally determined electron distributions terminate at 12–13 eV, i.e., below the threshold (ϵ_i = 15.5 eV) of the direct ionization process, it is not possible to use these distributions in the calculation of the direct ionization frequency z on the basis of Eq. (3.16). We can extrapolate the distribution curves to the ionization region using the calculations given in [58] because the calculated and experimental curves agree quite well in the subthreshold region (Fig. 18). As pointed out earlier, the value of the reduced field E/N governs entirely the distribution function. For E/N = 1.3 · 10⁻¹⁵ V · cm² the calculated value of the averaged (over the velocity distribution) direct ionization cross section $\langle \sigma_i v\rangle_{calc}$ is ~0.4 · 10⁻¹² cm³/sec, which is almost an order of magnitude smaller (Table 1) than the experimentally determined value $\langle \sigma_i v\rangle_{exp}$. At 2 Torr this discrepancy between the experimental $\langle \sigma_i v\rangle_{exp}$ and calculated values is even greater. The discrepancy is clearly due to multistage (cascade) ionization [91]. This is supported by the observation that an increase in the current causes the ratio $\langle \sigma_i v\rangle_{exp} / \langle \sigma_i v\rangle_{calc}$ to increase by a factor of 2 when the current is increased from 10 to 60 mA.

The steady-state ionization balance equation (1.11) was solved by Spenke [92] for the case when the loss of charged particles was due to ambipolar diffusion and the ionization was a multistage process, i.e., $z = a + bN_e$, where a and b are constants. Spenke obtained the radial distribution of the electron density which was not of the Bessel type (a Bessel distribution was obtained only for direct ionization). The distribution was found to be narrower than the Bessel curve, i.e., the electron density in the direction toward the tube walls fell more rapidly than in the direct ionization case. We carried out relative measurements of the electron density along the tube radius in N_2 and N_2-Xe (4:1) discharges; the results are plotted in Fig. 21. Although the differences between the Schottky (a) and Spenke (b) results were not very large, the experimental points fitted better the Spenke curve. This provided an additional confirmation of the occurrence of multistage ionization in the positive column in our experiments.

Let us now consider which excited states of the N_2 molecule can participate in the multistage ionization process. The N_2 molecule has several metastable electronic states. For example, the triplet state $A^3\Sigma_u^+$ with an energy 6.1 eV is the lowest level in the first positive system and, according to [93], its radiative lifetime is about 1 sec. The state $a^1\Pi_g$ mentioned earlier (its energy is 8.25 eV), as well as the states $a^1\Sigma_u^-$ and $^3\Delta_u$ with energies between 7.78 and 8.28 eV, can — according to [94] — have nearly equilibrium populations governed by the effective electron temperature. If the states participating in the multistage ionization process and their populations are known, we can use our values of $\langle \sigma_i v\rangle_{exp}$ to find the multistage ionization cross section.

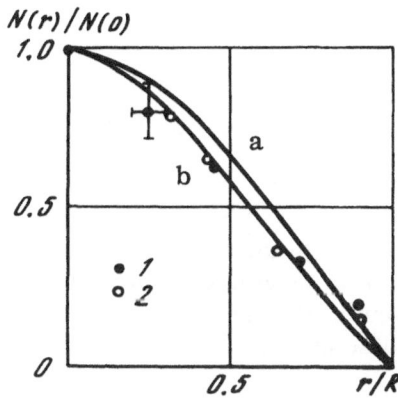

Fig. 21. Radial distribution of the electron density in a tube of 20 mm diameter containing N_2 at p = 1 and 2 Torr (1) and containing an N_2-Xe mixture at p = 1.5 Torr (2); i_d = 34 mA.

If direct ionization takes place and there is no gas heating, the dependence of the electric field on the current should not be parallel to the abscissa [95]. The multistage ionization process apparently reduces the effective ionization potential and this facilitates creation of new electrons needed to maintain the discharge current so that the electric field falls with rising current. It is possible that in our case the multistage ionization process is affected by the gas temperature and, in its own turn, it influences the dependence of the electric field on the current (Fig. 17).

§ 2. Discharges in CO_2 and in Mixtures of Other Gases with CO_2

The operation of a CO_2 laser is influenced strongly by the composition of the gas mixture and the rate of flow. It would be of great interest to determine the influence of the separate components of a mixture on the properties of the CO_2 laser plasma. We shall report here the results of our studies of the dependences of various electrical properties of discharges on the admixtures of N_2 and He and also on the rate of gas flow.

Discharges in CO_2

Although discharges in pure CO_2 are not being used for stimulated emission, we carried out measurements in pure CO_2 gas in order to determine the role of dissociation of CO_2 and of the influence of admixtures of N_2 and He. Figure 22 gives the electron energy distribution function $f'(eV)$ for a discharge in CO_2 enclosed in a tube of 20 mm diameter; the discharge current was i_d = 55 mA and the pressure was p = 2.4 Torr. For comparison, this figure includes also the corresponding distribution for pure nitrogen under the same discharge conditions. It is clear from Fig. 22 that the greatest differences between these distributions occur in the region of the tail, i.e., in CO_2 the number of electrons decreases more rapidly with rising

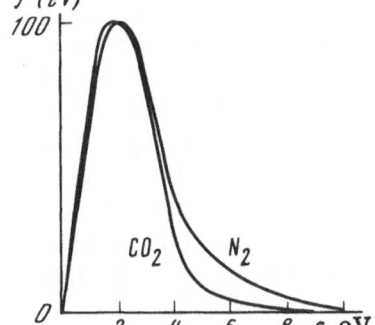

Fig. 22. Electron energy distribution $f'(eV) = c\sqrt{\nabla}f_0$ (eV).

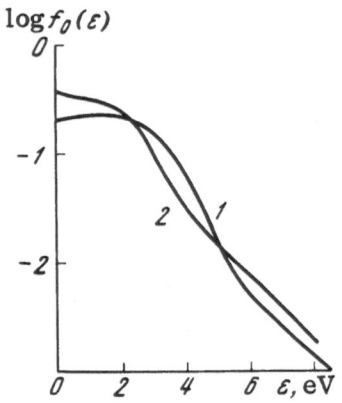

Fig. 23. Electron energy distribution in CO_2 discharges in a tube of 20 mm diameter (i_d = 55 mA, p = 2.4 Torr): 1) continuous-flow discharge; 2) static discharge.

energy than in N_2. In comparing these two curves we must bear in mind that they are normalized in such a way that the maximum electron energies are equal; actually, the total electron density in an N_2 discharge is almost twice as high as in CO_2. At low energies (ε = 0-3 eV) the distributions are practically identical. The curves in Fig. 22 apply to a continuous-flow discharge with a gas traveling at a velocity of about 1 m/sec.

The electron energy distribution in pure CO_2 under static conditions is quite different (Fig. 23). The average energy in the distribution decreases from 2.5 to 2.2 eV. The longitudinal electric field rises from 24 to 30 V/cm but the electron density (determined by the resonator method) is unaffected.

The changes observed when a continuous-flow discharge is replaced with a static one are due to the changes in the plasma composition caused by the dissociation of CO_2. It follows from the work of Ochkin [9] that the dissociation of CO_2 in a continuous-flow discharge depends on the flow velocity and on the discharge current: It decreases monotonically along the length of the tube. In our experiments the probe used to measure the distribution was located at a distance of 9 cm from the point of entry of the gas into the discharge zone. For this position of the probe and for the stated flow velocity and discharge current, the degree of dissociation at the point where the probe was located was less than 5%. However, in a static discharge this degree could reach 40-50%. Then, a large number of the CO molecules appeared in the discharge and these molecules were known to have large cross sections of inelastic collisions with electrons. For example, it was reported in [96] that the cross section for the excitation of vibrations in the CO molecule was $\sigma_{max} = 8 \cdot 10^{-16}$ cm^2 when the electron energy was ε = 2 eV, i.e., it was almost three times as high as for N_2. Naturally, the appearance of CO in the discharge gave rise to additional inelastic electron losses so that their average energy decreased. We applied the energy balance equation (3.7) to electrons in the positive column and found that the change from the continuous flow to a static system reduced the product $\delta_{eff} \nu_{eff}$ by a factor of almost 1.5.

The electron energy distribution functions were calculated in [82, 97] for CO_2 laser mixtures as a function of the ratios of the CO_2 and CO concentrations. The results of these calculations were in qualitative agreement with our measurements. They demonstrated that the appearance of CO in the mixture and a corresponding fall in the amount of CO_2 reduced the electron energy because of the loss of the number of electrons of energies exceeding 3 eV. When CO appeared in the mixture, the fraction of the electron energy dissipated in the excitation of the vibrational levels of the CO molecule increased. For example, when the degree of dissociation was γ = 0.3, this fraction was 15% for a $CO-CO_2-N_2-He$ (0.5 : 0.5 : 1 : 8) mixture. In a later investigation [98], Nighan calculated the average electron energies for pure CO_2, pure CO, and a CO_2-CO (1 : 1) mixture as a function of the parameter E/N. These calculations showed,

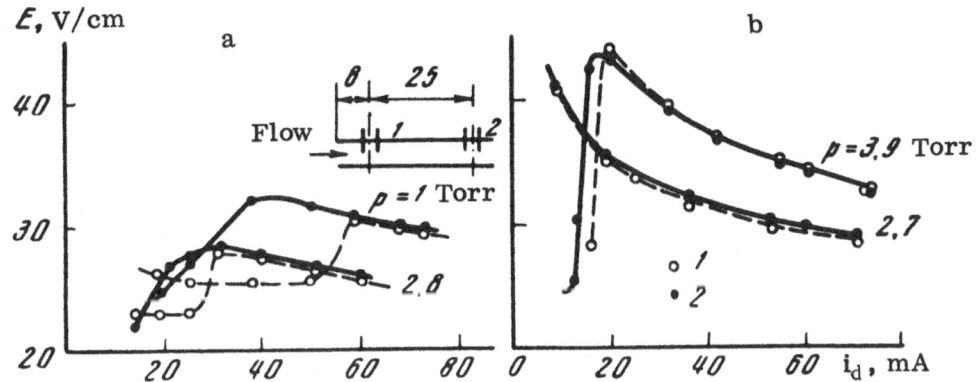

Fig. 24. Longitudinal electric field in CO_2 discharges in tubes of 20 mm (a) and 11 mm (b) diameters: 1) left-hand probe pair; 2) right-hand probe pair. The direction of flow is shown in the upper part of Fig. 24a.

in agreement with the experimental results, that the replacement of pure CO_2 with a mixture of CO_2 and CO reduced the average energy of electrons. However, one should remember that the calculations reported in [82, 97, 98] were carried out without allowance for the dissociation as an elementary process with a definite threshold energy and a definite probability but were simply carried out as a function of the composition of the mixture.

The influence of the dissociation of CO_2 explained also the results of measurements of the electric field. We soldered two pairs of probes into a discharge tube and measured the field intensities at distances of 8 and 25 cm from the point of entry of the gas into the discharge zone (Fig. 24a). Since the chemical composition of the dissociating flowing gas varied along the tube, the field intensities given by different pairs of probes could be different. Figure 24 gives the field intensities for pure CO_2 in tubes of 11 and 20 mm diameter. When the discharge current was large, the degree of dissociation was practically the same in different parts of the tube and the dependences of the field intensity on the discharge current obtained using the left-hand (1) and right-hand (2) pairs of probes were identical and had the usual falling form. When the discharge current was reduced, the field intensity first fell for the left-hand pair of probes and then for the right-hand pair; this was due to the fact that a steady-state degree of dissociation (under given discharge conditions) was not yet achieved at the position of the measuring probe pair. It should be pointed out that the relative distribution of the anomalies in the dependences of the field intensity on the discharge current (Fig. 24) reflected qualitatively the kinetics of the dissociation of CO_2 considered as a function of the pressure, discharge current, and tube diameter [9].

Discharge in CO_2-N_2 Mixtures

It is known that the addition of nitrogen increases considerably the output power of a CO_2 laser [1]. This occurs because of the efficient transfer of the vibrational energy from the N_2 molecules to the antisymmetric vibrations of the CO_2 molecules (the lowest antisymmetric vibrational level of CO_2 is the upper laser level). It would be interesting to investigate also the influence of nitrogen admixtures on the electrical properties of CO_2 laser plasmas. Unfortunately, it was difficult to record the second derivative because of strong noise and fluctuations in a CO_2-N_2 discharge. The interference was so strong that we were unable to determine the distribution in continuous-flow CO_2-N_2 discharges at any pressure, current, ratio of the components, or parameters of the external circuit.

However, we were able to determine the electron energy distribution in a sealed system when 20 min elapsed from the striking of a discharge in a tube with molybdenum electrodes.

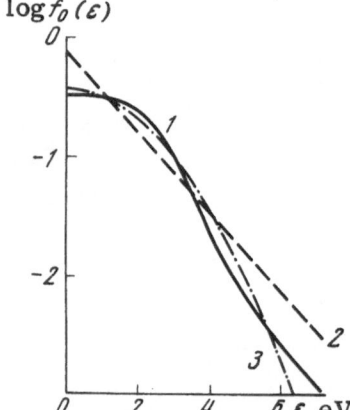

Fig. 25. Electron energy distribution in a CO_2-
N_2 (1 : 1) mixture under static conditions ($\bar{\varepsilon}$ =
2.0 eV): 1) experimental curve; 2) Maxwellian
distribution; 3) Druyvesteyn distribution.

This distribution is plotted in Fig. 25. The distribution in CO_2-N_2, like the distributions in
pure N_2 and CO_2, differed considerably from the Maxwellian and Druyvesteyn functions with
the same average electron energies. These average energies were usually lower than in pure
CO_2 and N_2 discharges under equivalent conditions. This reduction in the average electron
energy as a result of addition of CO_2 to N_2 could be explained by a reduction in the role of the
collisions of the second kind between electrons and vibrationally excited N_2 molecules. When
CO_2 was added to N_2, the vibrationally excited N_2 molecules transferred energy mainly to the
CO_2 molecules which was then emitted as radiation or lost by relaxation. This was confirmed
by measurements of the vibrational temperatures [78]. When CO_2 was added to N_2 the vibrational
temperature fell from 4000 to 1200°K and became comparable with the vibrational temperature
of the antisymmetric modes.

The electron density in a CO_2-N_2 mixture (Fig. 26) rises linearly with the current. It
should be stressed that, in contrast to discharges in pure N_2, the electron density is independent
of the pressure (within the limits of the experimental error). Figure 26 gives also the dependences of the ion current flowing to the tube walls. As in the pure N_2, this current increases
superlinearly with the discharge current. The ionization frequencies z_{exc} calculated on the
basis of these data are plotted in Fig. 27. The rise of z_{exc} with the discharge current shows
that the ionization process in CO_2-N_2 mixtures is of multistage type. On the whole, z_{exc} for
CO_2-N_2 mixtures is less than the corresponding values of z_{exc} for N_2. This confirms the observed difference between the electron distributions at high energies. The longitudinal electric fields in CO_2-N_2 mixtures are usually higher than in pure N_2 or CO_2 at the same pressures
(Figs. 17 and 24).

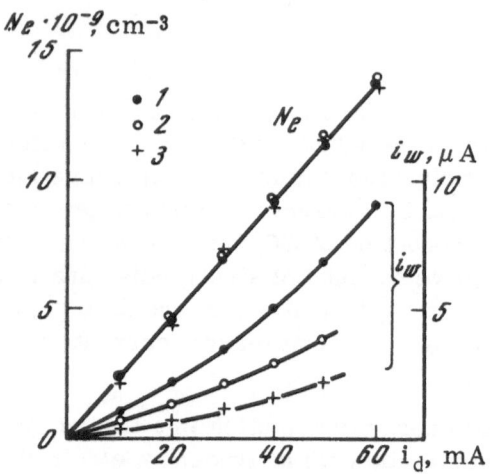

Fig. 26. Electron density N_e and ion current to
the tube walls in a CO_2-N_2 (1 : 1) mixture in a
tube of 20 mm diameter at p = 1, 2, 3 Torr
(curves 1-3, respectively).

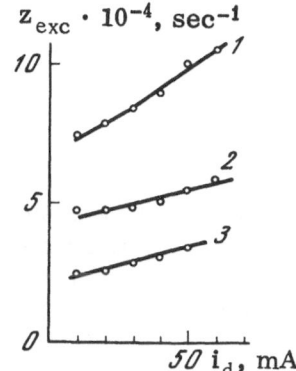

Fig. 27. Ionization frequency z_{exc} in a CO_2-N_2 (1:1) mixture enclosed in a tube of 20 mm diameter at p = 1, 2, 3, Torr (curves 1-3, respectively).

Discharges in CO_2-N_2-He Mixtures

The output power of carbon dioxide lasers is usually increased by addition of He in amounts several times greater than the amounts of CO_2 and N_2. It is shown in [74, 75] that the addition of He to CO_2 laser mixtures reduces the gas temperature because of the high thermal conductivity of helium. Moreover, helium increases the rate of relaxation of the lower laser level and reduces the population of this level so that the population inversion becomes greater. It is interesting to find how electrical properties of discharges are affected by the addition of He.

We shall first consider the influence of He on the electron energy distribution function. Figure 28 shows the distribution functions for a CO_2-N_2-He (1:1:4) mixture at a pressure p = 2 Torr and different discharge currents $i_d = 23$, 41, and 59 mA. The observed distributions differ from the Maxwellian and Druyvesteyn functions when compared on the assumption of the same average electron energy. They have the same characteristic features as in mixtures free of He, in spite of the fact that the amount of helium in the mixture is greater than that of molecular gases. This is to be expected because He does not scatter electrons inelastically in the energy range 0-19.8 eV and does not affect the distribution function, i.e., it acts as a buffer gas. However, there are some differences. If we compare the distributions obtained for CO_2-N_2 and CO_2-N_2-He mixtures at the same total pressure, we find that the distribution in the ternary mixture with He is characterized by higher energies and this is mainly due to the tail in the electron distribution. The result is fully expected because when molecular gases with large inelastic scattering cross sections are replaced with an inert gas (helium) the total losses experienced by electrons decrease. The fact that $\delta_{eff}\nu_{eff}$ for a mixture with He is half the corresponding product for $CO-N_2$ confirms this conclusion.

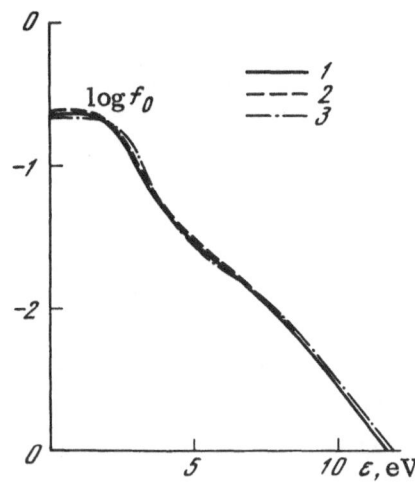

Fig. 28. Electron energy distribution in a CO_2-N_2-He (1:1:4) mixture at p = 2 Torr (continuous-flow conditions, $E/N = 0.8 \times 10^{-15}$ V · cm^2): 1) $i_d = 59$ mA, $\bar{\varepsilon} = 2$ eV; 2) $i_d = 41$ mA, $\bar{\varepsilon} = 3.1$ eV; 3) $i_d = 23$ mA, $\bar{\varepsilon} = 2.95$ eV.

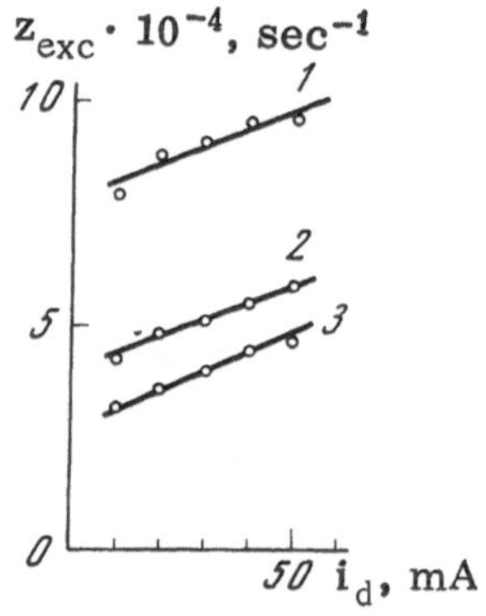

Fig. 29. The ionization frequency z_{exc} in a CO_2-
N_2-He $(1:1:4)$ mixture enclosed in a tube of
20 mm diameter and calculated using Eq. (3.12)
for $p = 2$ Torr (1), 3.1 Torr (2), and 42 Torr (3).

The higher energy of electrons in mixtures containing He is reflected also in the measured ionization frequencies (Fig. 29). These frequencies are usually higher than the corresponding values for CO_2-N_2 mixtures (compare with Fig. 27) and for pure N_2 (compare with Table 1). The dependence of z_{exc} on the discharge current is not destroyed by the addition of He and this shows that the ionization in the positive column of a mixture containing He is of multistage type.

The electron density (Fig. 30) in a ternary mixture varies linearly with the discharge current. The electron density in a mixture with He does not differ from the density in a binary mixture or in pure nitrogen. An increase in the ionization frequency z_{exc} on addition of He is clearly compensated by a reduction in the ambipolar diffusion time.

The influence of discharge conditions on the electron distribution in the tail can be determined by investigating the behavior of the intensities of bands in the second positive system of N_2.

When the discharge current is increased, the intensity of these bands increases linearly in the same way as in pure N_2 and this intensity follows the total electron density in the discharge. This shows that the distribution function is almost independent of the discharge current.

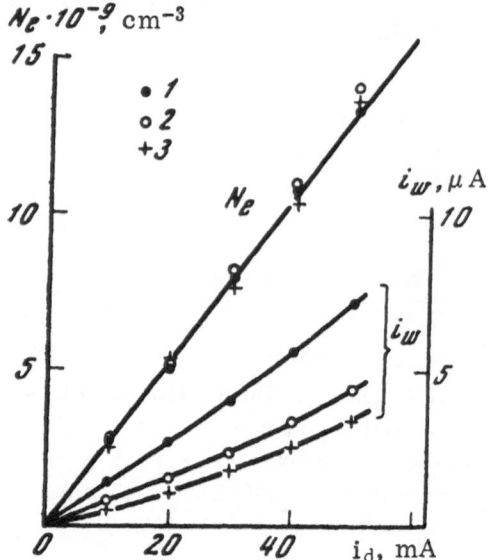

Fig. 30. Electron density in a CO_2-N_2-He
$(1:1:4)$ mixture at $p = 2$ Torr (1), 3.1 Torr (2),
and 4.2 Torr (3).

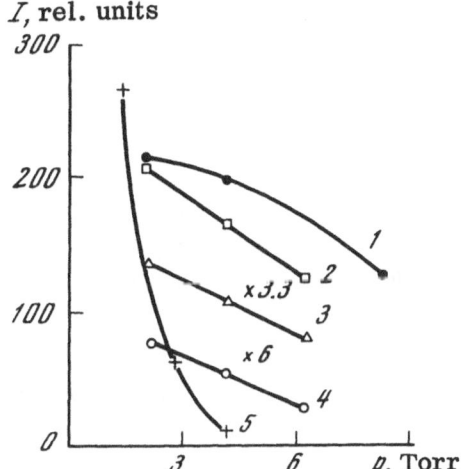

Fig. 31. Dependence of the intensity of the (0, 2) band of the 2^+ system of N_2 on the pressure in a tube of 20 mm diameter (i_d = 20 mA) in different gas mixtures: 1) CO_2-N_2-He (1:3:8); 2) CO_2-N_2 (1:2); 3) CO_2-N_2 (2:1); 4) CO_2-N_2 (9:1); 5) N_2.

When the gas pressure is increased (Fig. 31), the intensity of the 2^+ bands of N_2 decreases to a much smaller degree than in pure N_2. This behavior is due to a change in the tail of the distribution function. A comparison of the values of I/p_{N_2}, where I is the integrated intensity of the bands emitted from CO_2-N_2 (1:2) and CO_2-N_2-He (1:3:8) mixtures at the same total pressure, shows that the number of electrons with energies in excess of 11-12 eV in a ternary mixture is twice as high as this number in helium-free mixtures. Similar behavior of the intensities of the 2^+ bands of N_2 are reported in [80]. The band intensities decrease monotonically when the pressure is increased from 0.05 to 0.9 Torr. The addition of He causes the band intensities to rise by a factor of 1.5-2, but when the discharge current is increased, this rise becomes smaller. Unfortunately, it is not possible to compare directly the results reported in [80] with those in the present paper because the pressures in [80] are lower than in our case and the electric fields are not given.

Our experiments [10] show that the electron density in ternary mixtures depends linearly on the discharge current density, which is also true of molecular gases. High-frequency measurements in tubes of different diameter (20, 18, 12, and 6 mm) have shown that the electron density is independent of the tube diameter but is governed by the discharge current density. In all the investigated mixtures (apart from pure N_2) the electron density corresponding to a given discharge current density remains constant (within the limits of the experimental error) when the pressure is increased from 0.5 to 6 Torr and it rises linearly with the discharge current density. For example, in the case of a discharge in a CO_2-N_2-He (1:1:4) mixture in a 6-mm tube the value of N_e rises from $1.2 \cdot 10^9$ to $4 \cdot 10^{10}$ cm^{-3} when the current density is increased from 3 to 100 mA/cm^2. At higher discharge current densities the linear rise of the electron density is no longer observed. The dependence of N_e (cm^{-3}) on the discharge current density can be described by the approximate relationship

$$N_e = 4.03 \cdot 10^8 j_d - 1.13 \cdot 10^3 j_d^3, \qquad (3.17)$$

where j_d is the discharge current density (mA/cm^2). In the case of discharges in CO_2, the deviation from linearity is observed much earlier, in the current density range j_d = 15-18 mA/cm^2.

§ 3. Influence of Xe on Properties of CO_2

Laser Plasmas

The first brief report of a study of the influence of Xe on the properties of CO_2 lasers was given by Paananen [99], who found that the laser efficiency increased by 12-15% when small

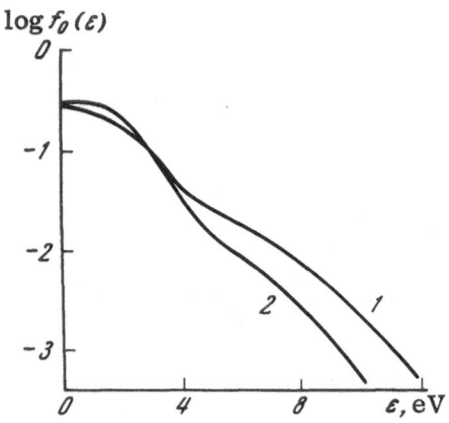

Fig. 32. Electron energy distribution functions
for different mixtures: 1) CO_2-N_2-He (1:1:6),
p = 4 Torr, $\bar{\varepsilon}$ = 3.1 eV; 2) $CO_2-N_2-He-Xe$
(1:1:6:1), p = 4.5 Torr, $\bar{\varepsilon}$ = 2.5 eV.

amounts of Xe were added to a CO_2-N_2-He mixture. Later and more detailed studies [30, 33]
of the influence of Xe on the properties of CO_2 laser plasmas confirmed this result. However,
the discharge plasmas were studied by these authors employing the usual one- and two-probe
methods applicable only in the case of a Maxwellian electron distribution. We demonstrated
earlier that such a distribution is inapplicable to electrons in CO_2 laser plasmas and, therefore, it was not surprising that the results reported in [30, 33] differed strongly (by a factor
of nearly 2) in respect of the absolute values of the electron temperatures (see § 5).

In the present section we shall give the results of our studies [100-101] of the influence
of Xe on the electron density and energy distribution, current to a probe, density of the ion
current to the tube walls, intensity of the longitudinal electric field, and intensity of the 2^+
bands of nitrogen. All the measurements were carried out in a sealed system. We studied
the influence of xenon on discharges in pure N_2 and O_2 as well as on discharges in N_2-CO_2 and
CO_2-N_2-He mixtures. In all cases the influence of Xe could be described as follows. The
addition of xenon reduced the average electron energy, mainly because of a reduction in the
number of electrons with energies in excess of 3-4 eV, but the distributions remained non-
Maxwellian.

Figure 32 shows the electron energy distributiom function for a CO_2-N_2-He (1:1:6)
mixture at p = 4 Torr and for the same mixture but with the addition of Xe. We can clearly
see that the addition of Xe reduces the number of fast electrons and increases the number in
the range of 0-2 eV. Garscadden and Bletzinger [31] measured the second derivative of the
probe current in a mixture with Xe and without it. They found that when Xe was added, the

Fig. 33. Dependences of the intensities of the
N_2 band on the Xe pressure: 1) CO_2-N_2-He
(1 : 1 : 4), p = 4 Torr, 2^+ system; 2) N_2, p = 1.4
Torr, 2^+ system; 3) CO_2-N_2 (1 : 1), p = 1.4 Torr,
2^+ system; 4) N_2, p = 1.4 Torr, 1^+ system.

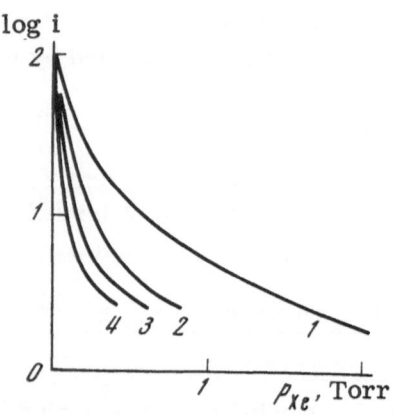

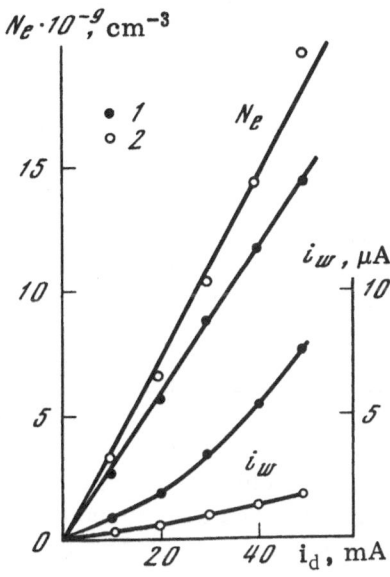

Fig. 34. Electron density N_e and ion current to the walls i_w in a tube of 20 mm diameter (static conditions): 1) CO_2-N_2-He (1 : 1 : 4), $p = 2.2$ Torr; 2) $CO_2-N_2-He-Xe$ (1 : 1 : 4 : 1), $p = 2.6$ Torr.

number of fast electrons in the tail decreased but on the whole the average energy increased because of the shift of the second-derivative maximum in the direction of higher energies.

The addition of Xe altered the color of the discharge. Measurements of the intensities of the 1^+ and 2^+ bands of N_2 indicated that these intensities fell strongly even in the presence of a small amount of Xe (Fig. 33) both in the case of pure N_2 and in the case of N_2-CO_2 mixtures. Since the excitation of the $C^3\Pi_u$ state of the nitrogen molecule, from which the 2^+ system began, was due to direct electron impact from the ground state $X^1\Sigma_g^+$ [78], the behavior of the intensities of the 2^+ bands system reflected changes in the electron distribution in the region of the excitation threshold (11-12 eV). The intensity of the 2^+ bands decreased by almost one order of magnitude on addition of 0.5 Torr Xe to the ternary mixture, which was in qualitative agreement with the measurements of the electron energy distribution (Fig. 32).

The fall of the 2^+ band intensities in pure N_2 and in CO_2-N_2 mixtures was stronger than in ternary mixtures. It was interesting to note (curves 2 and 4 in Fig. 33) that the intensity of the 1^+ bands, characterized by a lower excitation threshold ($\varepsilon_i = 7.25$ eV) than the 2^+ system, was affected to a greater degree. Moreover, no traces of the 1^+ system were found in CO_2-N_2 mixtures. This was clearly due to the fact that the $B^3\Pi_g$ state, from which the 1^+ system began, had a radiative lifetime of the order of 10^{-5} sec [102], i.e., a lifetime three orders of magnitude longer than that of the $C^3\Pi_u$ state. Therefore, the deactivation of the $B^3\Pi_g$ state was not only due to radiative transitions but also due to quenching collisions with, for example, the CO_2 molecules. In the presence of Xe in the mixture, these quenching collisions could be due to Xe atoms. This resonant energy transfer from N_2 to the radiation-emitting levels of the $6p_5-6p_{10}$ group of Xe was observed in the afterglow emitted by $Xe-N_2$ mixtures [103].

Addition of Xe increased the total number of electrons in a discharge. This was confirmed by microwave resonator measurements and by measurements of the electron current to a probe at the vacuum potential. It is clear from Fig. 34, showing the discharge-current dependences of the electron density in a ternary CO_2-N_2-He (1 : 1 : 4) mixture at $p = 2.2$ Torr and for the same mixture but with the addition of $p = 0.4$ Torr Xe, that the electron density increased by 20-25% and the ion current to the tube walls decreased severalfold. The ioniza-tion coefficients z_{exc} deduced from the data in this figure were plotted in Fig. 35.

In all cases which we investigated the addition of Xe to a mixture or a pure gas reduced the longitudinal electric field. As pointed out earlier, the establishment of steady-state condi-

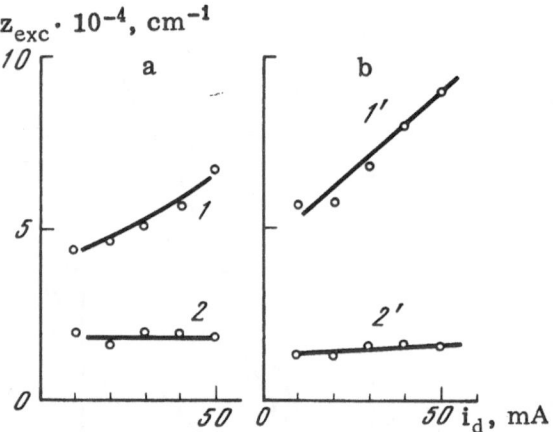

Fig. 35. Dependence of the ionization frequency z_{exc} on the discharge current i_d [Eq. (3.12), static conditions]: 1) CO_2-N_2 (1 : 1), p = 2 Torr; 2) CO_2-N_2-Xe (1 : 1 : 0.4); 1') CO_2-N_2-He (1 : 1 : 4), p = 2 Torr; 2') $CO_2-N_2-He-Xe$ (1 : 1 : 4 : 0.4).

tions in a discharge took 5–7 min [101]. Figure 36 shows the discharge-current dependences of the electric field in a ternary mixture with Xe and without it. The difference between field intensities shown in this figure ranges from 10 to 20%, depending on the discharge current.

Measurements of the ion current reaching a cylindrical probe demonstrated that the addition of Xe reduced strongly this current. Since the ion current depended on the ion mass, this reduction indicated that the main ion in discharges with Xe was Xe^+, whose mass was much greater than of any molecular ion present in the mixture. Changes in the ion composition of the plasma altered also the ambipolar diffusion conditions in the positive column. Since the mass of the Xe ion was much greater than that of other molecules, the mobility of this ion was less and, consequently, the diffusion slowed down. This was reflected in the ionization coefficients z_{exc} (Fig. 35) calculated from the balance equation (3.12) and from the measured values of the ion current to the tube walls.

The strong influence of small amounts of xenon on the properties of the plasma in a sealed CO_2 laser was mainly due to the fact that the ionization potential of Xe (12.1 eV) was 2–3 eV lower than the ionization potentials of the molecular components of the plasma. The appearance of a more easily ionizable component naturally facilitated creation of new electrons needed to maintain the discharge current. Therefore, for a constant current, the longitudinal field decreased and the average electron energy fell. The change in the nature of ions in the discharge did not alter the nature of the ionization process in the positive column. Estimates based on the experimental electron energy distribution indicated that the average cross section for the direct ionization of Xe, $\langle \sigma_i v \rangle_{calc}$, was $0.2 \cdot 10^{-12}$ cm^3/sec, i.e., it was 10 times smaller than the observed values of $\langle \sigma_i v \rangle_{exc}$. It was very likely that metastable $6s_5$ and $6s_3$ states with energies of 8.3 and 9.4 eV, respectively, participated in the multistage ionization of Xe. This process would require electrons of about 3–4 eV energy, i.e., with an energy fairly close to the average value.

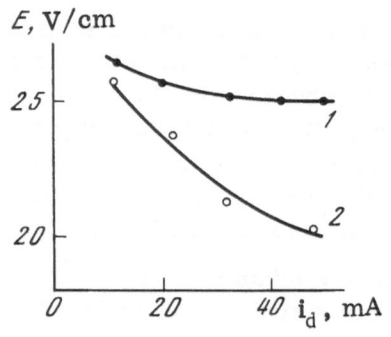

Fig. 36. Electric field in a tube of 20 mm diameter: 1) CO_2-N_2-He (1 : 1 : 4), p = 2.2 Torr; 2) $CO_2-N_2-He-Xe$ (1 : 1 : 4 : 0.4).

TABLE 2

Mixture	$N_2 (v=1-8)$	$CO (v=1-8)$	$CO_2 (00^01)$	$\bar{\varepsilon}$, eV
CO_2-N_2-He (1:1:6)	0.41	1.45	0.79	3.1
$CO_2-N_2-He-Xe$ (1:1:6:1)	0.48	1.70	0.93	2.5

Our experiments indicated that the addition of 0.5 Torr Xe to a CO_2-N_2 (1:1) mixture at a pressure p = 1.5 Torr increased the laser output power by a factor of 2. The output of a CO_2-N_2-He (1:1:4) mixture at p = 2.6 Torr was twice as high as that of the corresponding binary mixture and the addition of Xe in amounts between 0.5 and 1 Torr produced a further increase by a factor of 3. This increase in the output power could easily be understood bearing in mind the following points.

The upper laser level of CO_2 may be pumped by the transfer of energy from vibrationally excited N_2 and CO molecules, as well as by direct electron impact. Table 2 lists the averaged (over the electron velocities) excitation cross sections ($\langle \sigma_{vib} v \rangle \cdot 10^8$ cm^3/sec) representing the rate of pumping of the first eight vibrational levels of N_2 and CO and of the 00^01 level of the CO_2 molecule.

We can see from Table 2 that the addition of Xe to a molecular mixture increases the rates of pumping by 25-30%. Moreover, the output power is increased by an overall increase in the electron density in the discharge.

In considering the influence of Xe on the output power we must bear in mind also the influence of this additive on the gas temperature. However, it should be noted that whereas the influence of He on the gas temperature is direct (it is due to the high thermal conductivity of helium), the influence of Xe is indirect: The addition of this element reduces the longitudinal electric field. It is shown in [104, 105] that the gas temperature (more exactly, the temperature drop between the axis and the walls of the discharge tube) is proportional to the electrical power supplied per unit length of the discharge tube. It is clear from Fig. 36 that the addition of Xe is accompanied by a considerable reduction in the field intensity (for a given value of the discharge current) and consequently, of the power input. This reduction in the gas temperature should lower the rate of relaxation of energy stored in the antisymmetric vibrations of the CO_2 molecule and speed up the rate of exchange of the vibrational energy between the N_2 and CO_2 molecules. According to [106], the rates of these processes depend strongly on the gas temperature. Unfortunately, it is not possible to carry out a full quantitative analysis of the influence of Xe on the population inversion in CO_2 lasers because of the lack of information on the gas temperatures in mixtures containing Xe.

§ 4. Electrical Properties of Discharge

Plasmas and Population Inversion in CO_2 Lasers

Our experimental results must be compared with those obtained during laser action. The most important aspect is the existence of mixtures with certain ratios of the components which ensure the optimal conditions for the laser action in a given range of pressures and discharge currents.

We shall first consider the influence of nitrogen on the properties of CO_2 lasers. It is clear from [107] that the addition of nitrogen to CO_2 in comparable amounts increases the output power by a factor of 3-4. Our measurements of the electron energy distribution function indicate that the addition of N_2 results in narrowing of the electron energy distribution and a reduction in the average energy in such a way that the rates of excitation of the vibrational levels of the N_2, CO, and CO_2 molecules increase by just 10-15%. The total electron density in binary $CO_2 - N_2$ mixtures is 1.5-2 times higher than in pure CO_2. It is clear from these results that although qualitatively the changes in the parameters of the electron component of the plasma are in agreement with changes in the output power of a laser, we are far from a full quantitative explanation of the rise in the output power.

The main cause of the favorable influence of nitrogen lies in the fact that the vibrations of the nitrogen molecules are readily excited by electron impact and these vibrations are in resonance with the antisymmetric vibrations of CO_2 [1]. In spite of the fact that the rates of excitation of vibrations of the nitrogen and CO_2 molecules are comparable (Table 2), the addition of nitrogen results in an improvement of the population inversion. This is due to the following factors:

1) the rate of relaxation of the energy of the antisymmetric CO_2 vibrations as a result of collisions with N_2 is 2-3 times less than in the case of collisions with CO_2;

2) the replacement of some of the CO_2 molecules with N_2 reduces somewhat the gas temperature because of the higher thermal conductivity of nitrogen;

3) when vibrations are excited, it is very likely that several (up to eight) quanta are generated simultaneously and this results in a greater amount of energy stored in nitrogen for similar values of $\langle \sigma_{vib} v \rangle$ for N_2 and CO_2.

Let us now consider how He affects $CO_2 - N_2$ laser mixtures. Numerous experiments [108], including our observations, have demonstrated that the addition of He increases the output power by a factor of up to 10. It follows from our measurements of the distribution functions that the rates of pumping of the vibrational levels of the CO_2, N_2, and CO molecules do not increase and the total electron energy remains unaffected. Clearly, the favorable influence of He on CO_2 laser discharges results from the reduction of the gas temperature in the discharge and from more efficient depopulation of the lower laser level.

According to measurements reported in [74], the addition of He to $CO_2 - N_2$ mixtures in amounts usually employed in lasers reduces the gas temperature from 900-1000 to 400-500°K. When the temperature is reduced by this amount, the rate of vibrational relaxation of the upper laser level $00^0 1$ of CO_2 in collisions with CO_2 is reduced by at least one order of magnitude [109]. This reduction in the rate of vibrational relaxation increases the population of the upper laser level.

Lowering of the gas temperature reduces also the rate of deactivation of the deformation vibrations (lower laser level) by collisions with CO_2 molecules. However, this reduction may be fully compensated by collisions of CO_2 molecules with He atoms since He colliding with CO_2 deactivates much more effectively (by a factor of ~ 100) the lower laser level than the upper one.

Thus, two parameters are of major importance in the establishment of a population inversion: One is the electron density and the other is the gas temperature. It is found that the favorable effect of one of these factors (for example, an increase in the electron density as a result of an increase in the discharge current) reduces the role of the other factor (increase in the gas temperature). Such a simultaneous influence of two competing factors always gives rise to a maximum in the dependences of the output power of a laser on the discharge current.

It must be mentioned that the results of the present author's measurements of the electron densities and energy distributions [10, 11] were used in the author's laboratory [9] in calculations of population inversions in CO_2 lasers. These calculations were in good qualitative and quantitative agreement with the experimental determination of the laser gain [12].

Optimal conditions in gas-discharge CO_2 lasers correspond to an average electron energy of 2.5-2.7 eV and to a reduced field E/N = 3.5 · 10^{-16} V·cm². Calculations reported in [97] show that the strongest excitation of the vibrational levels of N_2, CO, and CO_2 participating directly or indirectly in population inversion is achieved at lower average electron energies ε ~ 1.2-1.3 eV. A reduction in the average electron energy should increase the rate of pumping of the upper laser level. It should be pointed out that in the case of optimal values of the parameter E/N the fraction of the electron energy acquired from the electric field and used to excite vibrational levels is approximately equal to the fraction used to excite electronic states. A reduction in the average energy would not only increase the rate of pumping of the vibrational levels but would also reduce the excitation of the electronic states of molecules which participate in the multistage (cascade) ionization processes. Therefore, the values quoted at the beginning of this paragraph are optimal for CO_2 laser action.

The pressures in optimal gas-discharge CO_2 lasers are usually higher (up to 10-15 Torr) than those investigated by the present author. Therefore, it would be interesting to know what changes would occur in the electron characteristics of the plasma as a result of an increase in the pressure. Our measurements indicate that the electron density N_e is practically independent of the gas pressure and is governed only by the discharge current density. An increase in the pressure by a factor of 2-3 should not alter greatly the behavior of the electron density in discharges.

The electron energy distribution and the average energy depend on the reduced field E/N. The parameter E/N decreases slowly with increasing pressure (Table 1) so that the average energy also decreases. However, we must bear in mind that when the pressure is increased, the gas temperature rises and it is this temperature that governs the population inversion.

At pressures p = 15-20 Torr the diffusion lifetime of electrons τ_d in a discharge column is comparable with the lifetime under bulk recombination conditions. This bulk recombination process may alter the radial distribution of electrons and may even cause contraction of the positive column.

§ 5. Probe and High-Frequency Measurements

in Non-Maxwellian Plasmas

Before the publication of calculations and measurements of the electron energy distribution, demonstrating the absence of Maxwellian distributions in molecular gases, there were several investigations [25, 26, 30, 33-37] carried out by methods valid only in the presence of a Maxwellian distribution. Qualitative dependences of the electron temperatures and band intensities on the discharge conditions obtained in these investigations are correct. However, the electron temperatures should be understood to be the effective values described by Eq. (3.4). Moreover, the absolute values of T_e should be treated with great caution. We shall now consider the investigations carried out under conditions closest to those used in the present study and find what results can be obtained by the use of ordinary probe measurements in a non-Maxwellian plasma.

The electron temperatures determined by the usual two-probe method [25, 26, 33] are very close to the effective temperatures of real distributions. It is well known that the two-probe characteristic is governed primarily by the high-energy electrons which are in the tail of the energy distribution. Our measurements of the distribution function have indicated that at energies of 7-10 eV the slope of the distribution function plotted on a logarithmic scale is close to the slope of a Maxwellian distribution with the same average electron energy (see, for example, Fig. 11). It is possible that it is this range of energies that dominates the contribution to the two-probe current—voltage characteristic. Consequently, the electron temperatures measured by the two-probe method should be very close to the effective temperature defined by $kT_{eff} = (2/3)\bar{\epsilon}$ and deduced from the real distribution. There is a quantitative as well as qualitative agreement (in the sense defined above) between the measurements reported in [25, 26, 33] and our results. For example, the electron temperature decreases with increasing total pressure in the mixture and also when nitrogen or xenon are added, but it is not greatly affected by the addition of helium.

In contrast to the two-probe characteristic, the one-probe results are governed by the main body of the electron energy distribution, i.e., by that part of the distribution where most of the electrons are located. If the electron temperature is deduced from that part of the current—voltage characteristic which is due to electrons of energies in the range 2-4 eV, the values of T_e may be understimated. (The part of the characteristic formed by electrons with energies up to 2 eV is easily assumed to be the electron saturation current. This is due to the electron sink effect, which slows down the rise of the electron current very strongly in this region.) The slope of the dependence log $f_0(\epsilon)$ in the 2 eV < ϵ < 4 eV range is usually 1.5-2 times less than for 7 eV < ϵ < 10 eV (Fig. 12). The electron temperatures deduced in this range are understimated in the same ratio. For example, the electron temperatures measured by the one-probe method in [30] in CO_2 laser discharges containing Xe are $T_e = 0.7$-1.4 eV and these values can be accounted for by the reason just stated.

Several measurements have been made of the "radiation temperatures" of CO_2 laser discharges [34-37]. By definition [67], the radiation temperature is

$$kT_r = -\int_0^\infty \frac{\nu(v)}{\nu^2(v) + (\omega - \omega_B)^2} f_0(v) v^4 dv \bigg/ \int_0^\infty \frac{\nu(v)}{\nu^2(v) + (\omega - \omega_B)^2} \frac{\partial f_0(v)}{\partial v} v^4 dv, \qquad (3.18)$$

i.e., it depends on the electron velocity distribution function and on the frequency of collisions between electrons and molecules; $\omega_B = eH/mc$ is the electron cyclotron frequency. If we analyze this expression with a distribution function of the $f_0(v) = \exp(-bv^y)$ form, we can show that the radiation temperature is equal to the electron temperature for a Maxwellian distribution (y = 2) or for any other distribution provided the collision frequency is independent of the velocity: $\nu(v) = $ const. In all other cases the value of kT_r is not equal to the effective temperature of the distribution kT_{eff}. If $\nu(v) \propto v$ (this dependence is exhibited approximately by the N_2 and CO molecules), the value of kT_r is less than $kT_{eff} = (2/3)\bar{\epsilon}$ and the deviation increases with the distribution parameter y. The application of a magnetic field makes it possible to determine the degree of departure of electrons from equilibrium (parameter y). For example, Noon et al. [34] studied nitrogen plasma under conditions close to our experiments and found that y = 6-7. For these values of y, the product kT_r is approximately 30% less than the corresponding product containing the effective temperature. This point must be borne in mind in comparison of our average electron energies with the results deduced from radiation temperatures. Typical values of kT_r given in [35-37] are 1.3-1.5 eV. If these values are corrected by 25-30% due to the non-Maxwellian nature of the distribution (y = 4-6), we find that our results are in satisfactory agreement (within 10%) with the measured radiation temperatures.

§ 6. Discharge Power Carried by Charges

to Tube Walls

The gas temperature in CO_2 laser discharges is one of the most important parameters that influence population inversion. In calculations of the gas temperature [104, 105] and in deriving the balance (rate) equations for the power supplied per unit length of the discharge column i_dE it is usual to assume that the power carried away by charges to the tube walls is negligible. This assumption can be checked if we know the density of the ion current flowing to the walls.

The proportion of the discharge power η_w converted to heat at the tube walls due to recombination and transfer of the kinetic energy of carriers to the walls is [95]

$$\eta_w^* = \frac{2\pi R j_w}{i_d E}\left(U_i + U_p + \frac{2kT_{\text{eff}}}{e}\right), \tag{3.19}$$

where U_i is the ionization potential of the molecule under investigation, U_p is the energy of the positive ions reaching the walls, and $2kT_{\text{eff}}/e$ is the average energy of the electrons reaching the walls. The energy U_p can be found from [94]

$$U_p \simeq \frac{kT_{\text{eff}}}{e} \ln \frac{j_{ew}}{j_w}, \tag{3.20}$$

where j_{ew} is the density of the random electron flux at the walls.

Thus, U_p can be assumed to be practically equal to the energy acquired by ions during the passage through the space-charge layer at the walls.

We shall find j_{ew} using the observation that in the diffusion regime in a positive column the ratio of the electron density at the walls N_{ew} to the electron density on the discharge axis N_{e0} is

$$N_{ew}/N_{e0} = \exp\left\{-\frac{e|V_r|}{kT_{\text{eff}}}\right\}, \tag{3.21}$$

where V_r is the difference between the potentials on the axis and on the walls. Then,

$$j_{ew} = \frac{N_{ew}\bar{v}_e e}{4}. \tag{3.22}$$

We shall obtain estimates using the maximum value of the ion current density to the walls found in our experiments: $j_w = 3 \cdot 10^{-2}$ mA/cm². We shall use typical values for the other parameters: $i_d = 60$ mA, $E = 25$ V/cm, $R = 1$ cm, $U_i = 15$ eV, $kT_{\text{eff}} = 2$ eV, $N_{e0} = 15 \cdot 10^9$ cm⁻³, $V_r = 5$ V. It then follows from Eqs. (3.19)–(3.22) that $\eta_w \leq 10^{-2}$. Then, the proportion of the discharge power lost due to the recombination of charged particles on the tube walls and because of the transfer of the kinetic energy of electrons and ions to the walls does not exceed 1%.

CONCLUSIONS

1. A detailed analysis was made of the electron sink effect produced by a cylindrical probe. It was found that at moderate pressures (p = 1–4 Torr) the sink effect may distort the probe-measured electron energy distribution. The degree of distortion depends on the pressure, probe radius, and electron energy. A method is described for deriving an unperturbed electron energy distribution function for the experimentally determined second derivative. The electron sink effect at a probe reduces the rise of the electron current, particularly when the probe potential is close to the vacuum potential. This effect perturbs also the average energy of electrons and their density deduced from the current flowing to the probe at the vacuum potential. Quantitative estimates were made of the distortions in the average energy and electron density.

2. The probe and microwave resonator methods were used in studies of discharges in nitrogen in a tube of 20 mm diameter at pressures $p = 1$–4 Torr using currents $i_d = 10$–70 mA. It was found that in all the investigated cases the electron energy distribution is not Maxwellian. The deviation of the distributions from the classical forms is due to the weak ionization and the considerable role of inelastic collisions between electrons and molecules, particularly those involving the excitation of the vibrational and electronic degrees of freedom and those causing ionization of the molecules. The average energies of the measured distributions are in the range 2.5–3.2 eV and are governed primarily by the reduced field E/N. The results of measurements of the distribution function are in satisfactory agreement with the results of calculations reported by other workers. The electron density in a nitrogen discharge increases linearly in the range $(2$–$15) \cdot 10^9$ cm^{-3} when the discharge current is increased in the range 10–70 mA. The electron density depends weakly on the pressure. The frequencies of collisions between electrons and molecules, measured by the resonator method, are $(2$–$8) \cdot 10^9$ sec^{-1} and they agree well with the frequencies calculated from the distribution function. The use of the ionization balance equation and measurements of the density of the ion current flowing to the tube walls indicate that the ionization in the positive column is mainly of the multistate type.

3. Investigations of the positive column of discharges in pure CO_2 and in mixtures containing CO_2 showed that the distribution of electron energies was again strongly non-Maxwellian. The example of discharges in pure CO_2 was used in a demonstration that an increase in the degree of dissociation of CO_2 molecules reduces the average energy of electrons because of the appearance of CO molecules in the mixture.

The addition of nitrogen to pure CO_2 reduces the average electron energy and increases the electron density by a factor of 1.5–2. The average electron energy in $CO_2 - N_2$ mixtures is in the range 2–2.6 eV. The addition of nitrogen does not alter the nature of the ionization process, which remains of the multistage type.

Electrons in discharges occurring in ternary mixtures containing He have higher average energies (2.5–3.3 eV) than in a binary mixture without He, other conditions being equal. As in the case of pure molecular gases, the electron density in the presence of helium varies linearly with the discharge current in the range 10–70 mA and is $(2.5$–$15) \cdot 10^9$ cm^{-3}. High-frequency measurements in tubes of different diameter (20, 18, 12, and 6 mm) demonstrate that the electron density in ternary mixtures is independent (to within the limits of the experimental error) of the tube diameter or of the total pressure in the range 0.5–6 Torr, but is governed by the discharge current density.

4. The addition of small amounts of Xe to a sealed system alters strongly the properties of CO_2 laser plasmas. The addition of Xe increases the electron density, reduces the average energy of electrons and the longitudinal electric fields, alters the ion composition of the plasma, and reduces the density of the ion current flowing to the walls of the discharge tube. These changes are due to the lower ionization potential of Xe compared with other components of the plasma and they explain qualitatively why the output power and efficiency of CO_2 lasers increase on addition of Xe.

5. Measurements of the electrical parameters of plasmas in mixtures containing N_2 were combined with relative measurements of the intensities of the 2$^+$ bands of nitrogen. The changes in the band intensities confirmed qualitatively the validity of the measurements of the electron density and energy distribution.

6. The electron densities and energy distributions obtained in the present study were used to calculate population inversions in the author's laboratory [9] using the methods suggested in [7]. The results of these calculations were found to be in satisfactory qualitative and quantitative agreement with the measurements based on the gain method [12].

7. Measurements of the properties of non-Maxwellian plasma carried out using the standard probe and high-frequency methods may give incorrect results. The examples of other investigations carried out under similar discharge conditions are used to show the type of error to be expected.

The author is deeply grateful to Professor N. N. Sobolev for directing this investigation and his constant interest. He is also grateful to A. G. Sviridov, A. I. Lukovnikov, and V. N. Ochkin for fruitful help.

LITERATURE CITED

1. N. N. Sobolev and V. V. Sokovikov, ZhETF Pis'ma Red., 4:303 (1966); 5:122 (1967); Usp. Fiz. Nauk, 91:425 (1967).
2. C. K. N. Patel, W. L. Faust, and R. A. McFarlane, Bull. Am. Phys. Soc., 9:500 (1964).
3. V. P. Tychinskii, Usp. Fiz. Nauk, 91:389 (1967).
4. C. K. N. Patel, Appl. Phys. Lett., 7:15 (1965); Phys. Rev. Lett., 13:617 (1964).
5. G. J. Schulz, Phys. Rev., 116:1141 (1959); 125:229 (1962); 135:A988 (1964).
6. J. D. Swift, Br. J. Appl. Phys., 16:837 (1965).
7. B. F. Gordiets, N. N. Sobolev, and L. A. Shelepin, Zh. Eksp. Teor. Fiz., 53:1822 (1967).
8. É. N. Lotkova, V. I. Makarov, L. S. Polak, and N. N. Sobolev, Khim. Vys. Energ., 2:278 (1968).
9. V. N. Ochkin, Tr. Fiz. Inst. Akad. Nauk SSSR, 78:3 (1974).
10. M. Z. Novgorodov, A. G. Sviridov, and N. N. Sobolev, ZhETF Pis'ma Red., 8:341 (1968); Preprint No. 32 [in Russian], Lebedev Physics Institute, Academy of Sciences of the USSR, Moscow (1969).
11. M. Z. Novgorodov, A. G. Sviridov, and N. N. Sobolev, Zh. Tekh. Fiz., 41:752 (1971); Preprint No. 47 [in Russian], Lebedev Physics Institute, Academy of Sciences of the USSR, Moscow (1970).
12. E. T. Antropov, I. A. Silin-Bekchurin, N. N. Sobolev, and V. V. Sokovikov, IEEE J. Quantum Electron., QE-4:790 (1968).
13. V. L. Ginzburg and A. V. Gurevich, Usp. Fiz. Nauk, 70:201 (1960).
14. Yu. B. Golubovskii, Yu. M. Kagan, and R. I. Lyagushchenko, Zh. Eksp. Teor. Fiz., 57:2222 (1969).
15. Yu. M. Kagan and R. I. Lyagushchenko, Zh. Tekh. Fiz., 31:445 (1961); 32:192 (1962); 34:821, 1873 (1964).
16. Yu. B. Golubovskii (Ju. B. Golubowsky), Yu. (Ju.) M. Kagan, R. L. Lyagushchenko (R. J. Ljagustschenko), and P. Michel, Beitr. Plasma Phys., 8:425, 445 (1968).
17. L. S. Frost and A. V. Phelps, Phys. Rev., 127:1621 (1962).
18. A. G. Engelhardt, A. V. Phelps, and C. G. Risk, Phys. Rev., 135:A1566 (1964).
19. R. D. Hake and A. V. Phelps, Phys. Rev., 158:70 (1967).
20. J. B. Thompson, Proc. R. Soc. Lond., 262:503 (1961).
21. A. Garscadden and P. Bletzinger, Phys. Lett. A, 27:203 (1968).
22. A. Garscadden and D. A. Lee, Phys. Lett. A, 24:431 (1967).
23. Yu. M. Kagan, V. M. Milenin, N. K. Mitrofanov, and V. P. Smirnov, Zh. Tekh. Fiz., 39:1985 (1969).
24. K. F. Bessonova, O. N. Oreshak, A. F. Stepanov, and V. A. Stepanov, Zh. Tekh. Fiz., 41:100 (1971).
25. P. O. Clark and M. R. Smith, Appl. Phys. Lett., 9:367 (1966).
26. A. I. Carswell and J. I. Wood, J. Appl. Phys., 38:3028 (1967).
27. J. D. Rigden and G. Moeller, IEEE J. Quantum Electron., QE-2:365 (1966).
28. G. Schiffner and F. Seifert, Proc. IEEE, 53:1657 (1965).
29. A. Garscadden and S. L. Adams, Proc. IEEE, 54:427 (1966).

30. P. Bletzinger and A. Garscadden, Appl. Phys. Lett., 12:289 (1968).

31. P. Bletzinger and A. Garscadden, Proc. Ninth Intern. Conf. on Phenomena in Ionized Gases, Bucharest, 1969, Contributed Papers, publ. by Editura Academiei RSR, Bucharest (1969), p. 250.

32. P. Bletzinger and A. Garscadden, Phys. Lett. A, 29:265 (1969).

33. P. O. Clark and J. Y. Wada, IEEE J. Quantum Electron., QE-4:263 (1968).

34. J. H. Noon, P. R. Blaszuk, and E. H. Holt, J. Appl. Phys., 39:9 (1968).

35. J. H. Noon, P. R. Blaszuk, and E. H. Holt, J. Appl. Phys., 39:5518 (1968).

36. D. C. Tyte, Electron. Lett., 5:447 (1969).

37. J. Polman and W. J. Witteman, IEEE J. Quantum Electron., QE-6:154 (1970).

38. Yu. M. Kagan and V. I. Perel', Usp. Fiz. Nauk, 81:409 (1963).

39. L. Schott, in: Plasma Diagnostics Methods [Russian translation], Mir, Moscow (1971).

40. O. V. Kozlov, Electrical Probe in Plasma [in Russian], Atomizdat, Moscow (1969).

41. M. J. Druyvesteyn, Z. Phys., 64:781 (1930).

42. Yu. M. Kagan, V. L. Fedorov, G. M. Malyshev, and L. A. Gavallas, Dokl. Akad. Nauk SSSR, 76:215 (1951).

43. G. R. Branner, E. M. Friar, and G. Medicus, Rev. Sci. Instrum., 34:231 (1963).

44. R. L. F. Boyd and N. D. Twiddy, Proc. R. Soc. A, 250:53 (1959).

45. G. M. Malyshev and V. L. Fedorov, Dokl. Akad. Nauk SSSR, 92:269 (1953).

46. S. C. M. Luijendijk and J. Van Eck, Physica (Utr.), 36:49 (1967).

47. H. W. Drawin, Preprint EUR-CEA-FC-383 (1966).

48. S. C. Brown, Basic Data of Plasma Physics, MIT Press, Cambridge, Mass. (1959).

49. I. P. Zapesochnyi and V. V. Skubenich, Opt. Spektrosk., 21:140 (1966).

50. D. Rapp and P. Englander-Golden, J. Chem. Phys., 43:1464 (1965).

51. R. H. Sloane and E. I. R. MacGregor, Philos. Mag., 18:193 (1934).

52. S. M. Call, Rev. Sci. Instrum., 36:850 (1965).

53. J. D. Swift, Proc. Phys. Soc. Lond., 79:697 (1962).

54. Yu. M. Kagan and V. I. Perel', Zh. Tekh. Fiz., 35:2069 (1965).

55. A. I. Lukovnikov and M. Z. Novgorodov, Kratk. Soobshch. Fiz., FIAN, No. 1, 27 (1971).

56. H. Rothman, Stud. Cercet. Fiz., 9:53 (1958).

57. I. G. Petrovskii, Lectures on the Theory of Integral Equations [in Russian], Fizmatgiz, Moscow (1965).

58. A. I. Lukovnikov, E. P. Fetisov, and E. S. Trekhov, Zh. Tekh. Fiz., 40:1916 (1970).

59. N. A. Vorob'eva, Yu. M. Kagan, and V. M. Milenin, Zh. Tekh. Fiz., 33:571 (1963).

60. I. M. Bronshtein, Izv. Akad. Nauk SSSR, Ser. Fiz., 22:441 (1958): I. M. Bronshtein and V. V. Roshchin, Zh. Tekh. Fiz., 28:2200, 2476 (1958).

61. A. N. Soldatov, V. M. Klimkin, I. I. Murav'ev, and Yu. N. Gulyaev, Izv. Vyssh. Uchebn. Zaved. Fiz., No. 6, 149 (1970).

62. V. M. Milenin, Zh. Tekh. Fiz., 41:831 (1971).

63. A. A. Zaitsev, M. Ya. Vasil'eva, and V. N. Mnev, Zh. Eksp. Teor. Fiz., 36:1590 (1959).

64. A. I. Lukovnikov and M. Z. Novgorodov, Zh. Tekh. Fiz., 41:2433 (1971).

65. C. V. Goodall and D. Smith, Plasma Phys., 10:249 (1968).

66. V. E. Golant, Zh. Tekh. Fiz., 30:1265 (1960).

67. V. E. Golant, Microwave Plasma Investigation Methods [in Russian], Nauka, Moscow (1968).

68. M. A. Heald and C. B. Wharton, Plasma Diagnostics with Microwaves, Wiley, New York (1965).

69. J. C. Slater, Microwave Electronics, Dover, New York (reprinted 1969).

70. S. J. Buchsbaum, L. Mower, and S. C. Brown, Phys. Fluids, 3:806 (1960).

71. A. I. Anisimov, N. I. Vinogradov, and V. E. Golant, Zh. Tekh. Fiz., 33:1370 (1963); A. I. Anisimov, V. N. Budnikov, N. I. Vinogradov, and V. E. Golant, Zh. Tekh. Fiz., 35:2042 (1965).

72.   F. Borgnis, Helv. Phys. Acta, 22:261 (1949).

73.   G. D. Burdun, Zh. Tekh. Fiz., 20:813 (1950).

74.   A. G. Sviridov, N. N. Sobolev, and G. G. Tselikov, ZhETF Pis'ma Red., 6:542 (1967).

75.   A. A. Mikaberidze, V. N. Ochkin, and É. N. Lotkova, Zh. Prikl. Spektrosk., 16:426 (1972).

76.   R. A. Young and G. A. St. John, J. Chem. Phys., 49:3505 (1968).

77.   M. Saporoschenko, Phys. Rev., 111:1550 (1958).

78.   M. Z. Novgorodov, V. N. Ochkin, and N. N. Sobolev, Zh. Tekh. Fiz., 40:1268 (1970).

79.   M. Jeunehomme and A. B. F. Duncan, J. Chem. Phys. 41:1692 (1964).

80.   V. A. Koshel'kov, V. P. Malakhov, V. I. Pugnin, A. F. Stepanov, and A. N. Tekuchev, Elektron. Tekh. Ser. 3, No. 26 (1968).

81.   S. P. Sychev, Izv. Vyssh. Uchebn. Zaved. Fiz., No. 6, 60 (1958).

82.   W. L. Nighan, Appl. Phys. Lett., 15:355 (1969).

83.   I. P. Zapesochnyi and V. V. Skubenich, Proc. Fifth Intern. Conf. on Physics of Electronic and Atomic Collisions, Leningrand, 1967, p. 570.

84.   G. Herzberg, Phys. Rev., 69:362 (1946).

85.   R. W. Nicholls, J. Chem. Phys., 20:1040 (1952).

86.   S. L. Chen and J. M. Goodings, J. Chem. Phys., 50:4335 (1969).

87.   J. S. Townsend and V. A. Bailey, Philos. Mag., 42:873 (1921).

88.   V. N. Soshnikov and E. S. Trekhov, Zh. Tekh. Fiz., 37:1414 (1967).

89.   A. H. von Engel, Ionized Gases, Clarendon Press, Oxford (1959).

90.   W. H. Kasner and M. A. Biondi, Bull. Am. Phys. Soc., 9:184 (1964).

91.   B. N. Klyarfel'd, Zh. Tekh. Fiz., 8:2012 (1938).

92.   E. Spenke, Z. Phys., 127:221 (1950).

93.   N. P. Carleton and O. Oldenberg, J. Chem. Phys., 36:3460 (1962); E. C. Zipf, Jr., J. Chem. Phys., 38:2034 (1963); W. L. Borst and E. C. Zipf, Jr., Phys. Rev. A, 3:979 (1971).

94.   C. Kenty, Phys. Rev., 126:1235 (1962); J. Chem. Phys., 35:2267 (1961).

95.   V. L. Granovskii, Electric Current In Gases [in Russian], Gostekhizdat, Moscow-Leningrad (1952).

96.   M. J. W. Boness and G. J. Schulz, Phys. Rev. Lett., 21:1031 (1968).

97.   W. L. Nighan and J. H. Bennett, Appl. Phys. Lett., 14:240 (1969).

98.   W. L. Nighan, Phys. Rev. A, 2:1989 (1970).

99.   R. A. Paananen, Proc. IEEE, 55:2035 (1967).

100. M. Z. Novgorodov, V. N. Ochkin, A. G. Svirdov, and N. N. Sobolev, Kratk. Soobshch. Fiz., No. 11, 36 (1970).

101. S. D. Manukyan, M. Z. Novgorodov, and A. G. Sviridov, Elektron. Tekh. Gazorazryad. Prib., No. 2(18), 12 (1970).

102. V. N. Soshnikov and E. S. Trekhov, Opt. Spektrosk., 22:872 (1967).

103. V. N. Chirkov and A. V. Yakovleva, Opt. Spektrosk., 28:441 (1970).

104. A. V. Eletskii, L. G. Mishchenko, and V. P. Tychinskii, Zh. Prikl. Spektrosk., 8:425 (1968).

105. A. V. Gorelik, Elektron. Tekh. Gazorazryad. Prib., No. 2(18), 15 (1970).

106. W. A. Rosser, A. D. Wood, and E. T. Gerry, IEEE J. Quantum Electron., QE-4:336 (1968).

107. T. J. Bridges and C. K. N. Patel, Appl. Phys. Lett., 7:244 (1965).

108. G. Moeller and J. D. Rigden, Appl. Phys. Lett., 7:274 (1965); 8:69 (1966).

109. K. F. Herzfeld, Discuss. Faraday Soc., No. 33, 22 (1962).